Spaceships at the Final Frontier

Building Star Trek® Models

Rick Jackson

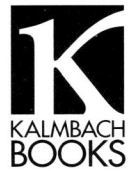
Kalmbach Books

© 2001 Rick Jackson. All rights reserved. This book may not be reproduced in part or in whole without written permission of the publisher, except in the case of brief quotations used in reviews. Published by Kalmbach Publishing Co., 21027 Crossroads Circle, Waukesha, WI 53187.

Printed in the United States of America

2000 01 02 03 04 05 06 07 08   10 9 8 7 6 5 4 3 2 1

Visit our website at
http://kalmbachbooks.com
Secure online ordering available

Publisher's Cataloging-in-Publication
(Provided by Quality Books, Inc.)

Jackson, Rick.
    Spaceships at the final frontier : building
Star Trek models / by Rick Jackson. — 1st ed.
    p. cm.
    ISBN: 0-89024-317-4

    1. Space ships—Models. 2. Star trek
(Television program) 3. Star trek films.
I. Title

TL844.J33 2000   629.47'022'8
                      QBI00-500067

Book design: Sabine Beaupré
Cover design: Kristi Ludwig

Contents

	Introduction: The Legend of *Star Trek* in Models	4
1	Construction 101: The Snap-fit *Star Trek* Spaceship Set	10
2	USS *Enterprise:* NCC-1701-E	17
3	The Klingon Bird of Prey Flight Display	25
4	Deep Space Nine	34
5	Lighting an *Ambassador*-Class Starship	44
6	Vacuum-formed Romulan Bird of Prey	60
7	The Cardassian *Galor*-Class Cruiser	75
	Photo Galaxy	81
8	The Borg Cube	92
Appendix 1	Paint	101
Appendix 2	Decals	102
Appendix 3	Tools, Supplies, Sources	103

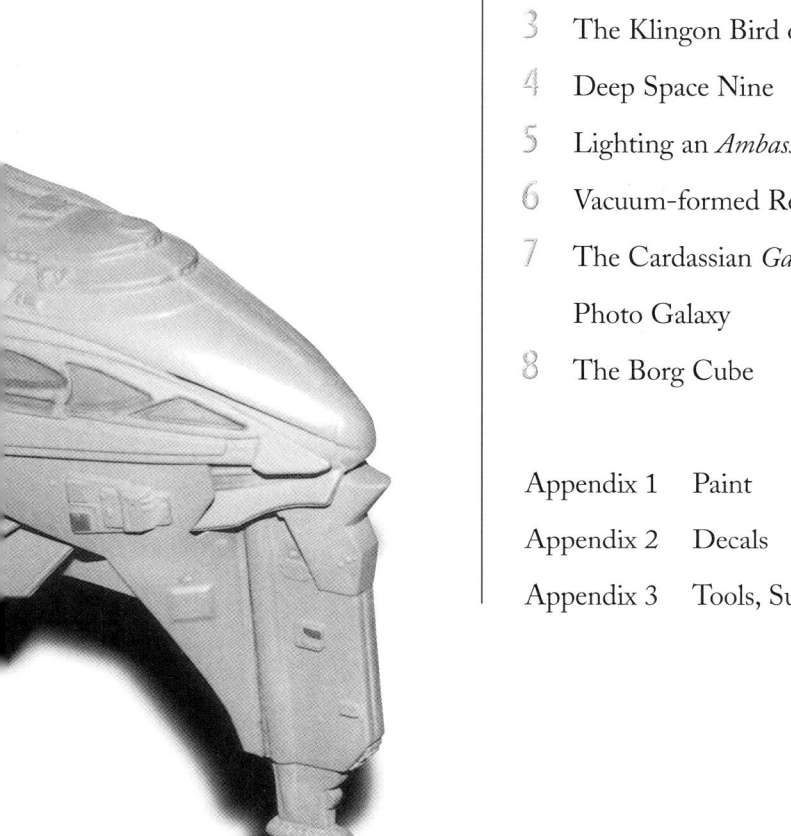

How significant has *Star Trek* become? This is a picture of the original production model taken at the National Air and Space Museum in 1987. The model has since been refurbished.

Introduction
The Legend of Star Trek in Models

Star Trek has become one of the largest and most enduring entertainment franchises in history. After a shaky start and almost near death on several occasions, interest in the story shows no real signs of tapering off, even after *Star Trek*'s 30th anniversary in 1997.

The development of *Star Trek* models parallels this evolution, from very simple beginnings to aftermarket products and books that can keep a modeler busy for years.

The early years – 1967-1978

From the original series (commonly referred to as "Classic Trek"), five kits were released by AMT (later to be AMT/Ertl). The *Enterprise* and the Klingon Battlecruiser were both released while the TV series was still in production. The Romulan Bird of Prey, *Galileo* shuttlecraft, and the K-7 Space Station (from "The Trouble with Tribbles") were all released in the mid-1970s.

AMT also released the Exploration Set consisting of a phaser, communicator, and tricorder. Aurora Models released a figure kit of Mr. Spock battling a snake in 1/12 scale. Also during this time, a model of the *Enterprise* bridge was released. This kit was not re-released until the 25th anniversary.

Aurora also had the rights to release the early kits of the *Enterprise*, the Klingon Battlecruiser, and Mr. Spock under their logo in the United Kingdom. Midori Models in Japan created an independent release of the *Enterprise* for their market during this time, since *Star Trek* was popular on Japanese television.

These early kits are highly sought after by kit collectors and can bring $100 or more. While the *Enterprise*, the *Galileo*, and the Klingon Battlecruiser have been re-released several times, making the kits affordable and easy to find, the Bird of Prey, the Exploration Set, and the K-7 in the large-scale formats have been released only once. When the first movie came out, the Mr. Spock figure was re-released with a movie-style uniform.

Light kits were a unique feature of the first releases of the *Enterprise* and the Klingon Battlecruiser. Both came with grain-of-wheat bulbs and mounting brackets for batteries to light the engine pods and bridge areas. Subsequent releases deleted these features.

The movies – 1979–1991

Star Trek appeared to be just another short-lived series until the early seventies. Two things happened to change that. First, reruns of episodes at all hours of the day built an audience that just couldn't get enough. The second event was the first *Star Trek* convention. Although only a few hundred people were anticipated, thousands showed up. During the early seventies, this interest gradually grew until Paramount decided to produce a movie reuniting the original series cast. In 1979, feeding off of the success of *Star Wars, Star Trek: The Motion Picture* premiered. While opinions differ on the quality of the story, there is no doubt that the special effects were a welcome improvement over those of the series. The *Enterprise* and the Klingon Battlecruiser received major facelifts to make them more impressive for the big screen. The movie also created a new vessel, a Vulcan shuttlecraft, which brought Mr. Spock to a rendezvous with the *Enterprise* early in the film.

The first movie was successful enough to warrant a sequel and, as they say, the rest is history. In all, six movies were made featuring the original cast. With each movie, the *Enterprise* and the Klingon Battlecruiser were re-released in new boxes with the current movie title (never mind that Klingons didn't even appear in some of the

movies). The Vulcan shuttle also was released as a kit and is now a collector's item.

While the movies seemed to appear like clockwork, some of the kits were slow to follow. In 1982, *Star Trek II: The Wrath of Khan* introduced the USS *Reliant*, a *Miranda*-class frigate, yet the *Reliant* kit wasn't released until 1995. In 1984, *Star Trek III: The Search for Spock* introduced both the USS *Excelsior* and the Klingon Bird of Prey. The Bird

of Prey and the *Excelsior* kits came out in 1994, most likely because they were featured frequently on *Star Trek: The Next Generation*.

The Next Generation – 1987–1994

What had been dismissed by many Hollywood executives as a waste of time turned into the proverbial 800-pound gorilla. The year 1987 brought us to the next installment of the legend: *Star Trek: The Next Generation*. Next Gen had a seven-year run, during which it consistently ranked as one of the top shows in syndication. The *Galaxy*-class *Enterprise-D*, fifth starship (and fourth type) to carry that name, was released in kit form in 1987. Other models, such as the Ferengi Marauder and the *D'deridex*-class Romulan Bird of Prey, appeared soon after they were seen on the show. They were packaged in a 3-in-1 kit with a small-format Klingon Bird of Prey. The larger *Vor'cha*-class Klingon Battlecruiser was released in 1991 as a single kit.

At this point kit production really hit its stride. For nearly ten years, there has been a constant supply of releases, including a healthy mix of new subjects to go along with the re-releases.

Star Trek: Deep Space Nine – 1993–1999

Imagine an 800-pound gorilla on steroids. That was *Star Trek* in the 1990s

In 1993 *Star Trek: Deep Space Nine* (DS9) premiered. Building upon a story line from Next Gen, the series introduced the planet Bajor, just coming out of a brutal occupation by the Cardassians. A Cardassian space station, Terek Nor, was occupied jointly by the Bajorans and the Federation and renamed Deep Space Nine. With the discovery of a stable wormhole near the planet, the station was moved near its mouth. DS9, once a backwater post, became a front-line center of attention.

As the series progressed, so did the kits. Early releases included the DS9 station in 1993 and the *Danube*-class Runabout shuttlecraft. The production staff felt that the series needed more flexibility, so they introduced a new threat. The Dominion and their race of soldiers, the Jem-Haddar, were from the Gamma Quadrant on the other side of the wormhole. Once the Runabouts were shown to be under-gunned to protect DS9, the USS *Defiant* was assigned to the station. The *Defiant* kit was released in 1996.

The Cardassians also used a variety of ships during the series. The primary one, the *Galor* class, was released as a kit in 1997.

Star Trek: Voyager – 1995–2001

Another spin-off from DS9/Next Gen, *Star Trek: Voyager* picks up the story as the Maquis, a group of (former) Federation citizens are waging a guerrilla war with the Cardassians. The Federation finds the Maquis tactics inappropriate, so they begin a cooperative campaign to stop them.

USS *Voyager* is a new *Intrepid*-class starship, smaller and more maneuverable than most Federation starships. This will allow it to pursue and combat the smaller Maquis ships. During her first assignment, *Voyager* is pulled into the Delta Quadrant along with some Maquis rebels. Faced with a 75-year journey home, they join forces to deal with the new races they encounter.

Concurrent with the release of the series, three kits were released in 1995. Breaking with tradition, Revell-Monogram was granted rights to produce the kits from the *Star Trek: Voyager* series. The initial release consisted of *Voyager*, the Maquis rebel ship shown in the first episode, and a Kazon (the original bad guys) ship. Since then, a Kazon torpedo and a small-scale 3-in-1 set have been released.

The Next Generation movies – 1994–?

In 1994 the baton was passed from the Classic cast to the Next Gen cast in the movie *Star Trek: Generations*. As the movie opens, Captain Kirk is participating in the launch and shakedown cruise of the *Enterprise-B*. Without recounting too much of the plot, the *Enterprise-D* is lost in a spectacular sequence, reminiscent of the destruction of the original movie *Enterprise*. The only new kit to come from this movie was the *Enterprise-B*, which is a modification of the *Excelsior* class. The *Enterprise-D* and the Klingon Bird of Prey were reissued in Generations box art.

Star Trek: First Contact introduces the fifth version of the *Enterprise*. The *Enterprise-E* is a *Sovereign*-class starship and is the largest yet. It incorporates some of the style features exhibited by *Voyager,* most notably the blended primary and secondary hulls.

As of this writing, the most recent movie is *Star Trek: Insurrection.* No kits have been released based upon the movie, although the *Enterprise-E* was released with new box art.

Variations on a theme

More recent releases have incorporated new features into older subjects to maintain the modeler's interest. Both the *Enterprise-D* and DS9 have been released with lighting kits. The lighting kit can also be purchased separately for use in other models. *Enterprise-D*, the Klingon Bird of Prey, and *Defiant* have been released in "Plus Packs," which contain paint, glue, and a brush in addition to the model.

The Cut-Away *Enterprise* was a 30th anniversary release. This is the Classic *Enterprise* in a scale similar to the movie *Enterprise* (although it claims to be nearly the same scale as the original kit release). Removable sections show the interior of the primary and secondary hulls and the warp engines.

Not only have the large-scale kits been released many times over the years, but the subjects have been released in other scales. An example is the Spaceship Set containing the Classic *Enterprise*, Classic Klingon Battlecruiser, and the Classic Romulan Bird of Prey. They are in a consistent scale of about 1/2200 and are (now) snap-fit kits. These were first released in 1976. Since then, we have also seen a 3-in-1 set of the Classic *Enterprise,* the movie *Enterprise,* and the *Enterprise-D*. Another 3-in-1 set has been released containing kits of the -C, -D, and -E.

The same subjects have also been recombined in other ways. A "scene" kit from the episode "The Enterprise Incident" consisting of two Romulan (Klingon) Battlecruisers and the Classic *Enterprise* has been released. The models mount onto a base with a sound chip that produces dialog clips from the show.

Additional kits

To this point I have focused exclusively on injection-molded kits from the major manufacturers. While this book emphasizes building these ships, they are by no means the only kits produced of *Star Trek* subjects.

Vacuum-formed kits and conversions were (and still are) quite popular. They are limited-run kits designed to cover subjects not yet available from the major manufacturers. For many years, the only source of an *Excelsior* model was a vacuum-formed kit released after the third movie. It is still a significant model, since it is designed to be in the same scale as the AMT/Ertl movie *Enterprise* and is more than twice the size of that kit.

Several conversions have also been produced to make the three-engined *Enterprise-D* from the series final episode "All Good Things . . ." and the Nebula-class starships, also starting with the *Enterprise-D.*

In smaller scales, there are numerous resin and white-metal kits. Subjects range from ships to star bases to even the Borg cube. The broadest selection of subjects would have to be the miniatures released for wargaming. There, you can see the entire gamut (official and unofficial) of Starfleet ships.

Boldly going where no one has gone before

These kits represent only the tip of the iceberg. Practically each episode or movie introduces a new ship or variant from either Starfleet or some race they have encountered. The possible subjects for models are nearly limitless. Many can be built using existing kits as a starting point. Others can be built by taking unrelated parts and creating a new ship. All you have to realize is that the process is the same whether you are in the studio or at your workbench.

Listing of Ship Kits Released by Major Model Companies

All kits are AMT or AMT/Ertl releases unless otherwise noted.
Multiple releases for different events (e.g. new movies) are indicated by a sequence number or event abbreviation.

I—*The Motion Picture*
II—*Wrath of Khan*
III—*Search for Spock*
IV—*Voyage Home*
V—*Final Frontier*
VI—*Undiscovered Country*
VII—*Generations*
VIII—*First Contact*

IX—*Insurrection*
FD—Flight Display kit
PP—Plus Pack including paint and brushes
RM—Revell/Monogram releases
FO—Fiber optic or light kit included
25—25th Anniversary release
30—30th Anniversary release

Subject	Kit Numbers	Original Release Date	Comments
USS Enterprise NCC-1701	S921 S951 6676	1967	Constitution class. Release S921 and some of the S951 releases had lights in engine nacelles and the top and bottom domes on the saucer. Almost uninterrupted production since early 1970s.
Klingon Battlecruiser	S952 6743(25)	1968 1991	D-7 class. Standard Klingon ship throughout the Classic TV series. First seen in the Classic episode "The Trouble with Tribbles."
Shuttlecraft Galileo	S959 6006 (25)	1974 1991	Marketed with both the Galileo and Galileo II decals. Introduced in the episode "The Menagerie Part I."
Enterprise Bridge	S950 6007(25)	1975 1991	
Romulan Bird of Prey	S957	1975	Bird of Prey for Classic TV series. Only seen in the episode "Balance of Terror."
Space Station K-7	S955	1976	Seen in the episode "The Trouble with Tribbles."
USS Enterprise NCC-1701 Klingon D-7 Battlecruiser Romulan Bird of Prey	S953 6677	1976 1989	3-in-1 set from the Classic series. All in 1/2200 scale. Enterprise in this kit is larger and assembles differently than the one in the 6618 3-in-1 set.
Klingon Battlecruiser	S971(I) 5111(II) 6682(III) 8229(VI)	1980 1982 1984 1991	D-7 class (but also called K'tan'ga class). Revamped version of Classic TV battlecruiser. First seen in *Star Trek: The Motion Picture* and in some later TV episodes.
USS Enterprise NCC-1701	S970(I) 5110(I) 6675(II) 6675(III)	1980 1982 1984	Revamped version of Classic TV starship. First seen in *Star Trek: The Motion Picture*. Destroyed in *Star Trek III: The Search for Spock*. Basic kit was re-released for each movie in a new box.
Vulcan Shuttle	S972(I) 5112(I) 6679(I) 6679(III)	1980	Used to bring Mr. Spock to the Enterprise in *Star Trek: The Motion Picture*.
USS Enterprise NCC-1701-A	6693 (IV) 6876 (V) 8617 (VI)	1986 1989 1991	Replacement for original Enterprise and introduced in *Star Trek IV: The Voyage Home*. Basic kit was re-released for each movie in a new box. Final kit version (for Star Trek VI) included a shuttlecraft and added some clear parts. A rarer version can also be found with a 1/2500 scale Enterprise-C.
USS Enterprise NCC-1701 USS Enterprise NCC-1701-A USS Enterprise NCC-1701-D	6618 6876 8617 8787(FD) 6005(25)	1988 1995 1991	3-in-1 set. All ships are in a consistent scale to each other but not the same scale as the first 3-in-1 set. The 25th Anniversary release is chrome plated.
USS Enterprise NCC-1701-D	6619 8793 8400 (PP) 8772 (FO)	1988 1996	Galaxy class. Seen throughout the series *Star Trek: The Next Generation*. Other ships in the class were also seen in that series and in *Star Trek: Deep Space Nine*. Destroyed in the movie *Star Trek: Generations*
Romulan Bird of Prey Ferengi Marauder Klingon Bird of Prey	6858	1989	Adversary 3-in-1 kit all in different scales. Romulan ship is D'deridex Class first seen in the Next Gen episode "The Neutral Zone." The Ferengi ship is first seen in the Next Gen episode "The Last Outpost."

Name	Kit #	Year	Notes
Klingon Battlecruiser	6812	1991	Vor'cha class. Infrequently seen Klingon ship. Introduced in the Next Gen episode "Reunion."
USS Enterprise NCC-1701-A	6957(25)	1991	Kit released with light kit and sound chip. Some parts, such as the main deflector, have been modified to enhance the lighting effect and to provide access for the light kit.
Runabout Rio Grande	8778 8764 8741	1993	Danube-class shuttlecraft. Primary shuttlecraft used in *Star Trek: Deep Space Nine*. First seen in the Next Gen episode "Timescape."
Deep Space Nine	8778 8764(FO)	1994	Originally thought to be in scale with the Enterprise-D from the 3-in-1 set. In fact, it is a much smaller scale.
Klingon Bird of Prey	8230(VII) 8015(PP) 6339(FD)	1994 1995 1997 1997	K'vort class. Introduced in movie *Star Trek III: The Search for Spock*. Has become the primary ship used by the Klingon Empire. Kit wings can be positioned in raised or lowered position (but are not movable).
USS Excelsior NCC-2000	6630	1994	Excelsior class. Introduced in *Star Trek III: The Search for Spock* (carrying an NX- designation). Other ships of this class have been seen frequently in the later TV series.
Kazon Warship	3606(RM)	1995	Predator class. The original bad guys on *Star Trek: Voyager*.
Maquis Ship	3605(RM)	1995	Peregrine class. Model is of version seen in the Voyager episode "Caretaker," but similar ships have been seen in Next Generation and DS9 episodes.
USS Enterprise NCC-1701	8790(30)	1995	Cut-away version in 1/500 scale, making it consistent with the movie Enterprise. Released in conjunction with the 30th anniversary.
USS Enterprise NCC-1701-B	8762	1995	Enhanced Excelsior class. Shown in the opening scenes of *Star Trek: Generations*.
USS Reliant NCC-1864	8766	1995	Miranda class. Seen in *Star Trek II: The Wrath of Khan* where it was destroyed. Other ships of this class (and variants) have been seen in the later TV series
USS Voyager NCC-74656 Kazon Warship Maquis Ship	8787(RM)	1995	3-in-1 set.
USS Voyager NCC-74656	3604(RM) 3612(RM	1995 1997	Intrepid class. Engines can be positioned for warp or sub-light speed. Limited-edition kit (3612) has a shuttlecraft hangar with shuttlecraft, an enhanced decal sheet, and different clear parts.
Kazon Torpedo	3608(RM)	1996	Special assault vehicle used in the *Star Trek: Voyager* episode "Maneuvers."
Legendary Space Encounters	8254(30)	1996	Classic Enterprise and Romulan Battlecruiser kits with fiber optics and a sound chip in the base depicting a scene from the episode "The Enterprise Incident."
Light Kit	6605 10DO	1996	Generic fiber-optic/bulb kit to add lights to a model.
USS Defiant NX-74205	8255 8398-10D0(PP)	1996 1997	Defiant class. Test bed for weapons to combat the Borg, it became the main long-range cruiser stationed at Deep Space Nine. Introduced in the episode "The Search, Part I."
Cardassian Cruiser	8324	1997	Galor class. Primary ship for the Cardassians. First seen in the Next Gen episode "The Wounded."
USS Enterprise NCC-1701-E	6326(VIII) 30065(IX)	1997 1998	Sovereign-class starship introduced in movie *Star Trek: First Contact*. Re-released for the movie *Star Trek: Insurrection*.
USS Enterprise NCC-1701-C USS Yamaguchi NCC-26510 USS Excalibur NCC-26517	8001D0 30038(FO)	1998 1999	Ambassador class. First seen in the Next Gen episode "Yesterday's Enterprise." The second release in clear plastic includes decals for the other ships plus a lighting kit.
USS Enterprise NCC-1701-B USS Enterprise NCC-1701-C USS Enterprise NCC-1701-E	8002D0	1999	Second variation of the 3-in-1 Enterprise theme. Again, in a consistent scale.

1

Construction 101: The Snap-fit Star Trek Spaceship Set

The Spaceship Set was the first of the small-scale, multi-ship kits to be released. It contains the *Enterprise*, a Klingon Battlecruiser, and a Romulan Bird of Prey, all from the original series. They are all in 1/2200 scale.

Because of their size, these might be the simplest *Star Trek* kits. The Bird of Prey has only three parts, while the *Enterprise* is the most complex at 12 pieces. Snap-fit kits use tight-fitting locating pins that rely on friction to hold the parts together. While they are marketed as entry-level kits for the youngest modelers, these kits are often good choices for builders of any age. For one thing, they may be the only kit of a particular subject or in a certain scale. Besides, the accuracy and detailing of the snap-fit kit are often equal to or better than the traditional kits.

Even though they are snap-fit, you will achieve the best appearance by using glue and filling the seams. With a few corrections, you will have a quick project that gives you nice models from the Classic Trek series.

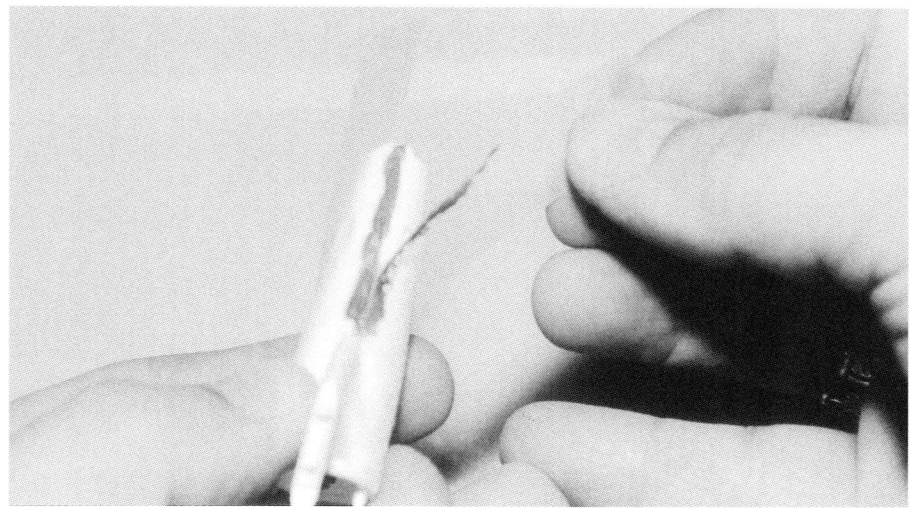

Fig. 1-1. Use tape to protect the areas around the seam when applying filler.

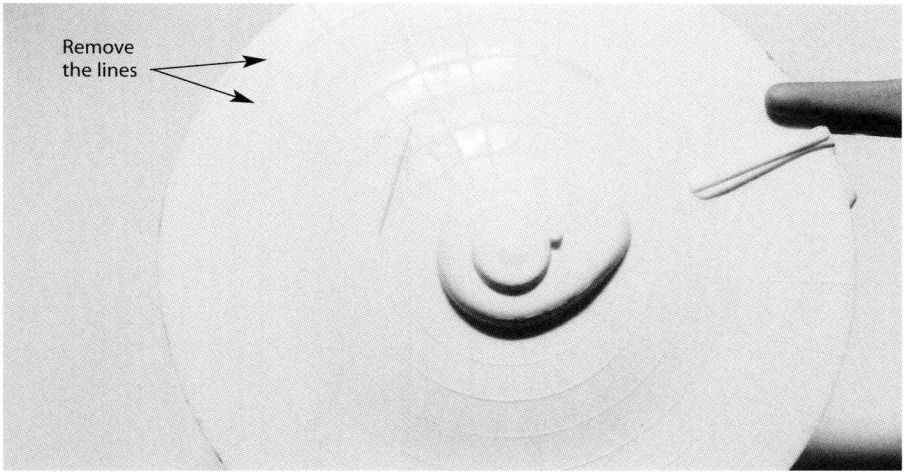

Fig. 1-2. Remove the panel lines from the top of the primary hull.

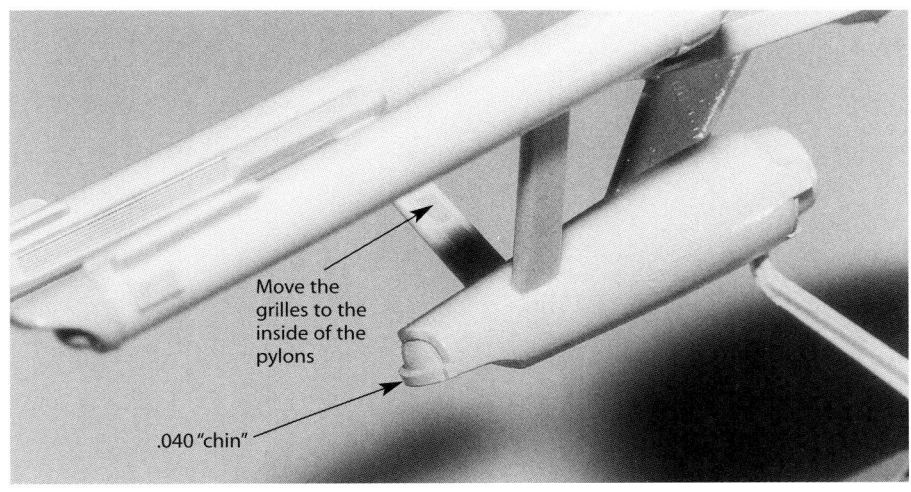

Fig. 1-3. Relocate the grilles on the engine pylons and add the "chin" to the shuttlecraft hangar.

Enterprise

When you are filling seams, use tape to protect the surface detail around the seam (fig. 1-1). Run a length of tape along each side of the seam. If the seam is large or deep, wait about a minute for the surface of the filler to "skin" dry to the touch and gently press on the filler with a wet finger to work it into the gap. Now peel back the tape. Rest your finger on the filler, if necessary, to prevent it from pulling up with the tape. If the tape does pull some of the filler away from the seam, push the filler back into place with the flat edge of your knife.

Remove the panel lines from the top of the primary hull (fig. 1-2). This is an error found on both this model and the large scale Enterprise. The lines are out of scale and should be replaced by light scribing or shading the paint. In this scale, leaving them off completely is just as effective.

When you look at the secondary hull from the side, the profile of the shuttlecraft hangar is too blunt. Add a "chin" of .040 strip styrene to extend the secondary hull to the proper length (fig. 1-3). Also, four grilles are molded on the outside of the engine pylons that should be on the inside. Sand off the kit grilles and add new ones with .005 sheet styrene.

You'll need to add a joint to the front of each nacelle. Use two to three layers of tape to create a straightedge and then scribe in the seam (fig. 1-4). Then use a needle file to enlarge it to the proper size.

Two coolant radiators are molded into the inboard pieces on each nacelle (part 7). The one closest to the seam is pyramid-shaped, making it too thick at the bottom (fig. 1-5). Cut them off and replace them with .020 styrene, 19mm long. Round the corners, making them similar to the remaining fins.

The edge of the main deflector dish is too thick. Sand down the face of the dish to thin it. This will result in a flat lip on the dish, so use an X-acto knife to blend it away (fig. 1-6).

Alignment of parts can be the trickiest part of building the *Enterprise* (fig. 1-7). If you're not careful, the engine nacelles may look OK but will be askew after the glue has dried. Remember that you are working in three dimensions. You can usually check any two directions at a time, so once the part looks right from one angle, check it from another.

Painting the *Enterprise* begins with a basic white scheme. For these models, I used—with a couple exceptions—the ModelMaster line from Testors. Dull it slightly with a light gray, such as Light Gray (FS36495) in a 50/50 mixture. A very pale blue-gray like Duck Egg Blue (FS 35622) or Russian MiG Blue (in the Xtracolor line) is appropriate for the rear ends of the nacelles, the recessed areas on the cooling vanes, and the radiators. Paint the main deflector Brass from the Metalizer line, also from Testors.

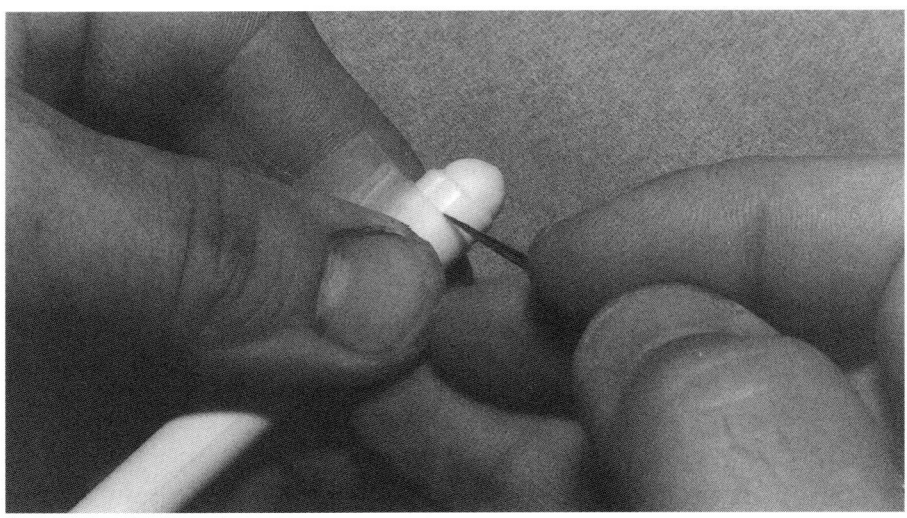

Fig. 1-4. Scribing the joint on the engine nacelle

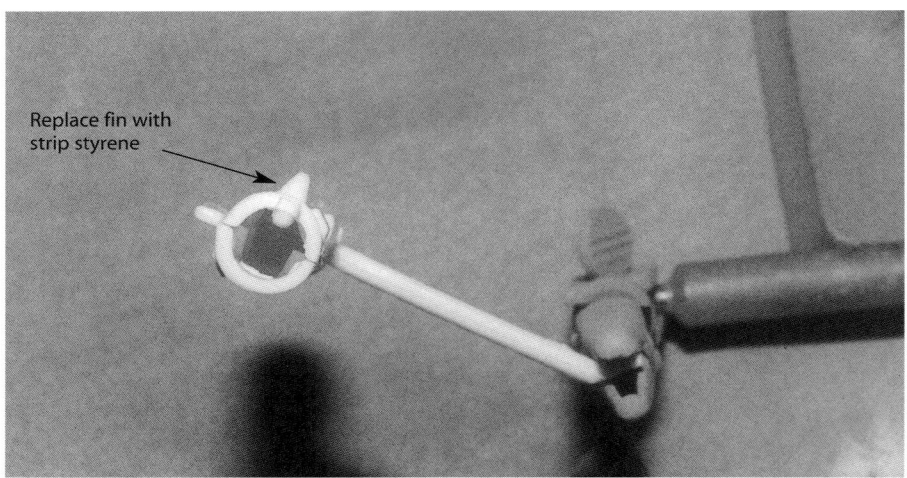

Fig. 1-5. The inboard cooling fin is too thick. Remove this one and replace it with strip styrene.

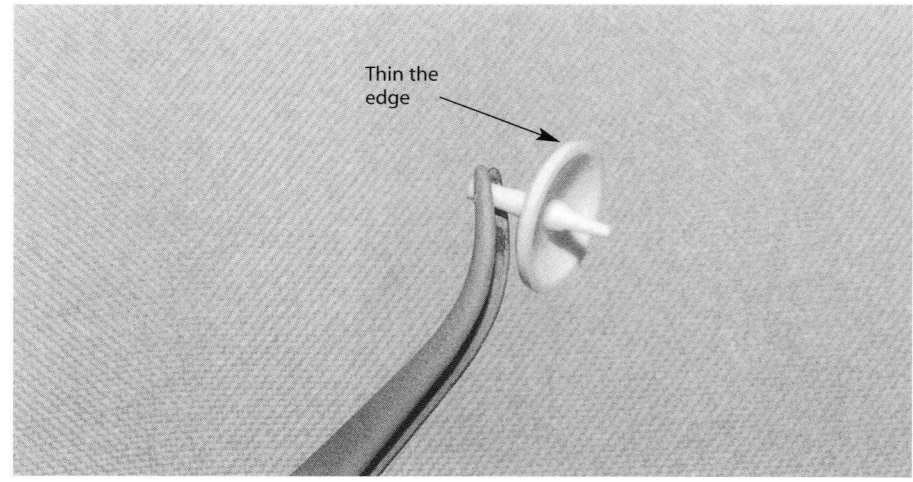

Fig. 1-6. Trim the edge of the main deflector dish to achieve the proper appearance.

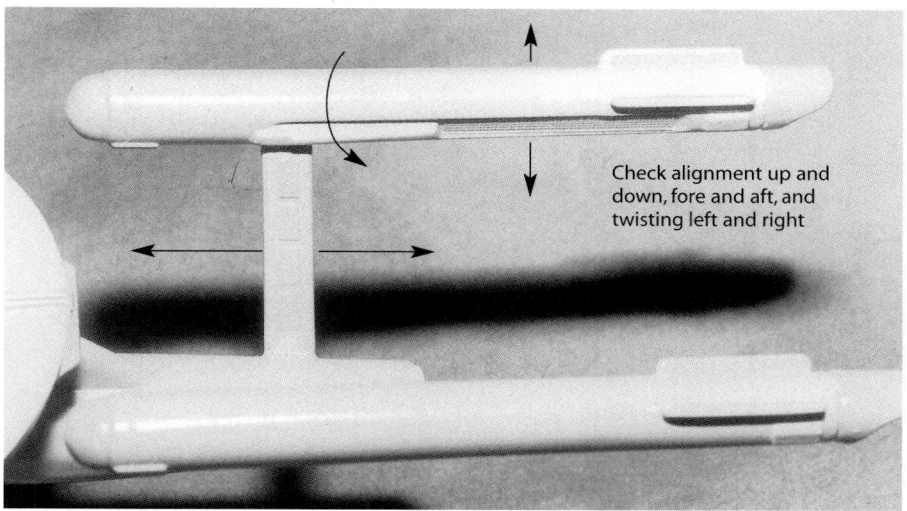

Fig. 1-7. Check the alignment of parts in three dimensions.

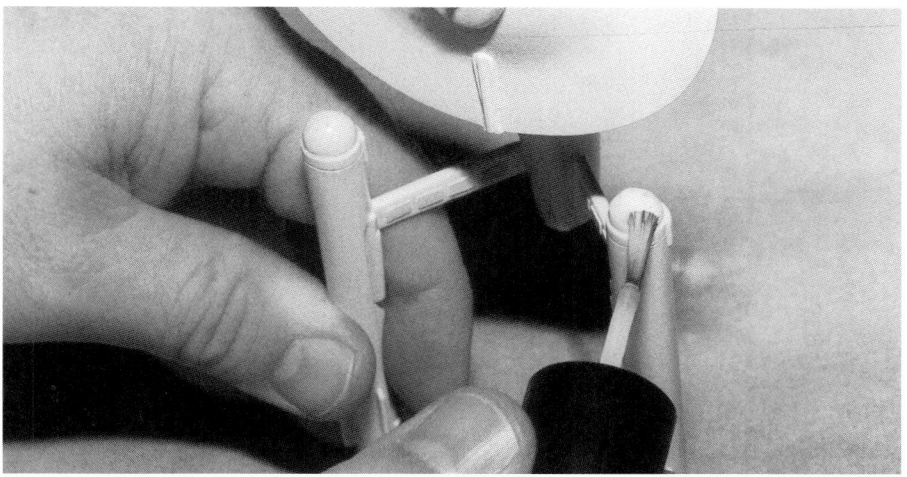

Fig. 1-8. Apply the off-white base coat of the nail polish.

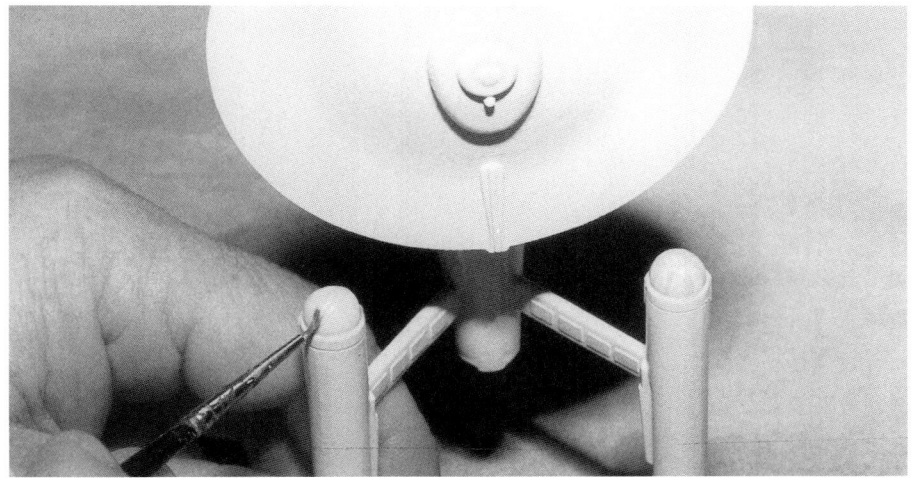

Fig. 1-9. Detail the nacelles with the gold highlights.

The nacelles also provide the biggest problem in painting. The kit is configured for the TV series, having a dome on the end of each nacelle and spinning gold lights at the front. If you want to build the engines with the simpler red front and smooth aft ends, which were seen only in the two pilots "The Cage" and "Where No Man Has Gone Before," some surgery is required. You must remove the domes from the rear of the nacelles and add small probes to the forward domes. This distinction is sometimes blurred in the episodes, since stock footage from the pilots was often used in the series for some orbit and fly-by shots.

So, how to accomplish the translucent, glowing effect? Go to the nearest cosmetics counter (incognito, if necessary) and look for nail polish! You want two colors—an opalescent off-white and a very pale gold with metallic highlights.

Start by applying the off-white to the nacelle fronts (fig. 1-8). Once each coat has been applied (this will probably take two to three coats), suspend the model with the painted surface down so the polish can gather on the point. Repeat this process for the domes on the aft end. Allow the polish 24 hours to dry completely.

Next, take the gold polish and, using a detail brush, gently apply a six-pointed pattern over the off-white (fig. 1-9). Use only a small amount on each coat, building up the intensity until you are satisfied.

Klingon Battlecruiser

Since these are snap-fit kits, the locating pins are oversized and the small size of the models causes the pins to extend through the model. This is a problem on the battlecruiser most of all, since the pin sockets stick out on the inside of the skinny engines (fig. 1-10). Once the nacelles are glued together, cut off the sockets and fill the area.

The undercut of the nose section is lost in the molding and must be restored (fig. 1-11). The starboard side is how the kit is molded, while the port side shows how the undercut has been restored using a triangular file and sandpaper. Also, use a needle file and sandpaper to reshape the photon launcher into a circular shape.

The photon torpedo launcher needs extra depth. Add a 2mm length of 1/8" tubing to the inside and blank it off with a piece of styrene. The neck is narrow and prone to breaking, so reinforce it with a piece of wire (fig. 1-12). Push some epoxy putty into the area around the bridge to support the area once it has been cleaned and sanded.

The Klingon Battlecruiser has a monochromatic paint scheme. The overall color is a medium gray, so use something in the Dark Ghost Gray (FS 36375) range.

Romulan Bird of Prey

The vertical fin of the BOP provides an opportunity for a modeling technique. Wherever practical, apply the glue to the joints on the inside of the model (fig. 1-13). You can use more if necessary with less risk of damaging the exterior detail. Liquid glue, applied with a paintbrush or drafting rule, will work its way around the joint through capillary action.

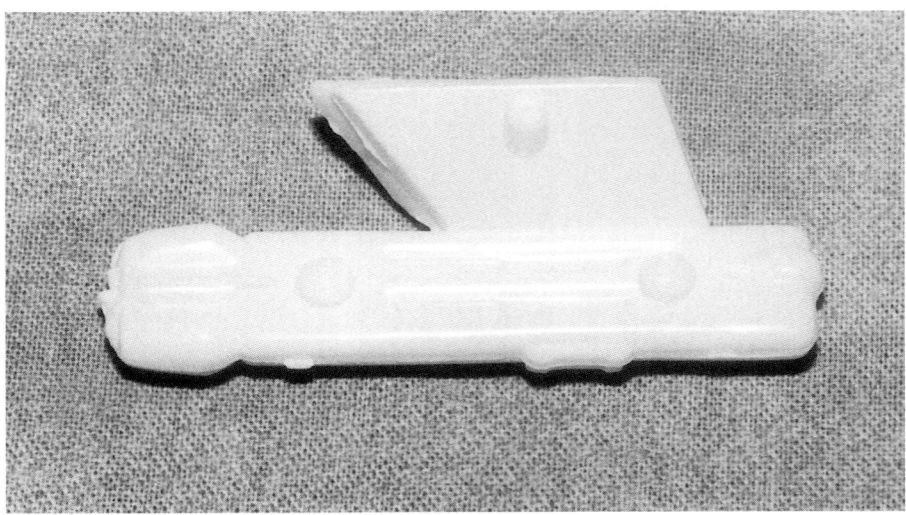

Fig. 1-10. Remove any visible sockets and pins. Fill and sand the area flat.

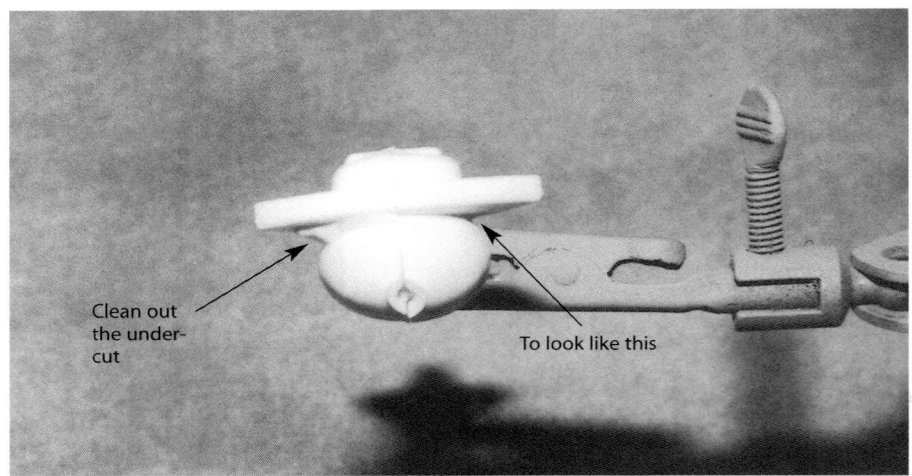

Fig. 1-11. Nose detail showing restored undercut (right)

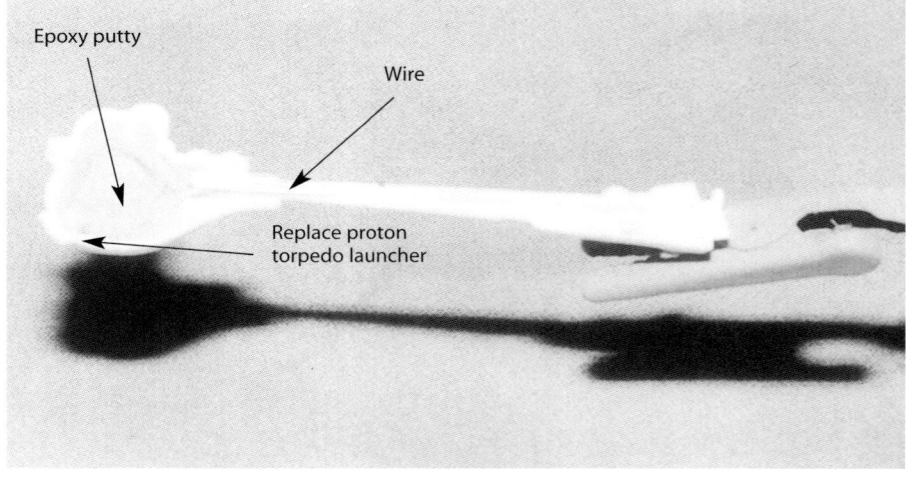

Fig. 1-12. Interior work on the Klingon battlecruiser

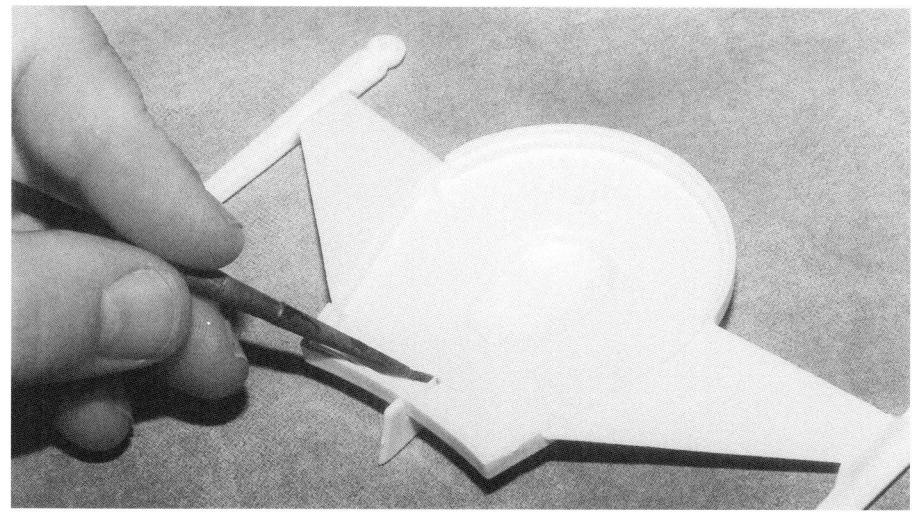

Fig. 1-13. Applying glue to a joint from the inside reduces the chance of damaging surface detail.

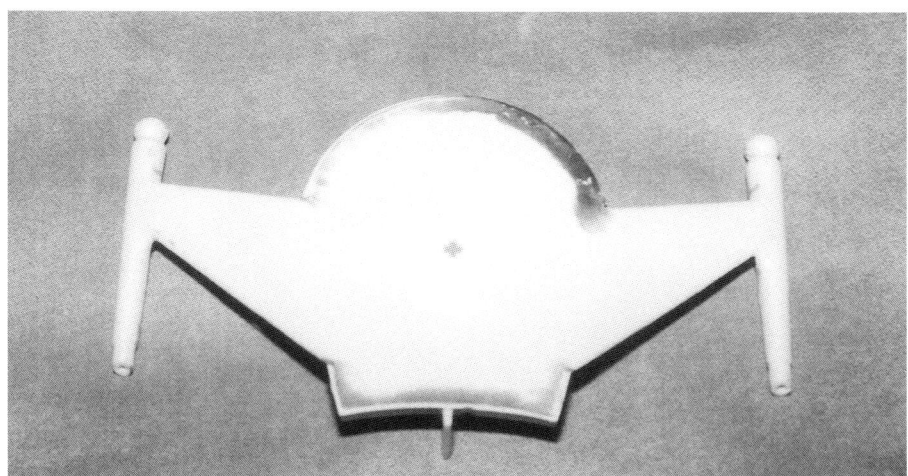

Fig. 1-14. Use a contrasting color for the filler wherever possible.

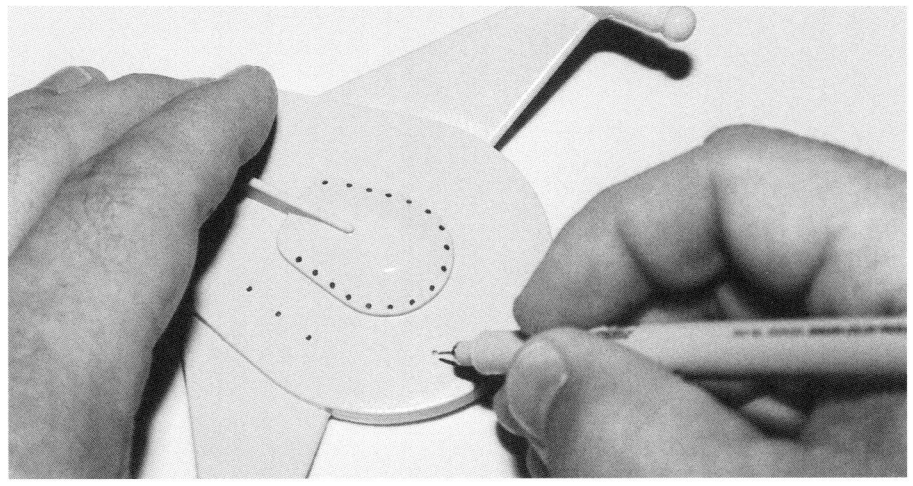

Fig. 1-15. The portholes and windows on the models can be added easily by using a .05mm black nylon-tipped pen.

Use a contrasting color for the filler wherever possible (fig. 1-14). This makes it easier to see how the seam has been filled and when to quit sanding. In this case, I have used green putty on white plastic. Other choices for the putty are white and, when using automotive putties, blue and red.

The Bird of Prey is also a rather nondescript gray. In the only episode to feature it, "Balance of Terror," it appears to have a slightly blue tint, so I chose to use the Duck Egg Blue again. This color also gives a little variety to the over-all display.

All three ships have windows that need attention. Hobby shops and drafting stores will have .05mm fine-tipped pens that can be used to fill the molded-in windows (fig. 1-15). Use the raised rim of the window frame as a guide and gently fill in the area. These pens use a water-based ink, so if you slip, you can wash off the mistake and start again. Just be sure to do this before applying the decal prep coat to protect the ink from the water used to apply the decals.

Decals

This kit illustrates one of the major problems with *Star Trek* kits. When building a model released before about 1980, you'll find that the decals that came with the kit are thick (fig. 1-16). Since then, the decals used are more like the ones modelers have become accustomed to—thinner and more flexible, willing to conform to the surface underneath.

Even though the locations for the decals are rather smooth, surface preparation is critical to avoid silvering. Refer to Appendix 2 for more on decal preparation.

There are no decal diagrams in the instructions. You are referred to the box art for placement. For the Klingon Battlecruiser and the Romulan Bird of Prey this is sufficient, but not for the *Enterprise*.

The decal sheet provides three sizes of letters and numbers, plus individual names for the ships mentioned in various episodes. See the chart to identify which NCC registry number goes with which name.

Registry No.	Ship
1017	Constellation
1700	Constitution
1701	Enterprise
1702	Farragut
1703	Lexington
1704	Yorktown
1705	Excalibur
1706	Exeter
1707	Hood
1708	Intrepid
1709	Valiant
1710	Kongo
1711	Potemkin

The ship name is centered across the top of the saucer 2mm from the front of the bridge structure. The largest set of registry numbers is also used on the top of the saucer. Even though they are printed on the decal sheet in pairs (NC C- 17 01), cut them apart and apply them individually. That way, they will form a smooth arc around the dish at the point where the central "hump" flattens out. This can't be done if they are left in pairs. The midsized letters go on the nacelles. How to use the smallest letters can't be determined from the box art. They are used on the underside of the saucer (fig. 1-17). Be careful to have them turned the right direction for the side they're on.

Fig. 1-16. The decals are probably the weakest part of this kit. They are thick with a lot of clear film around the image.

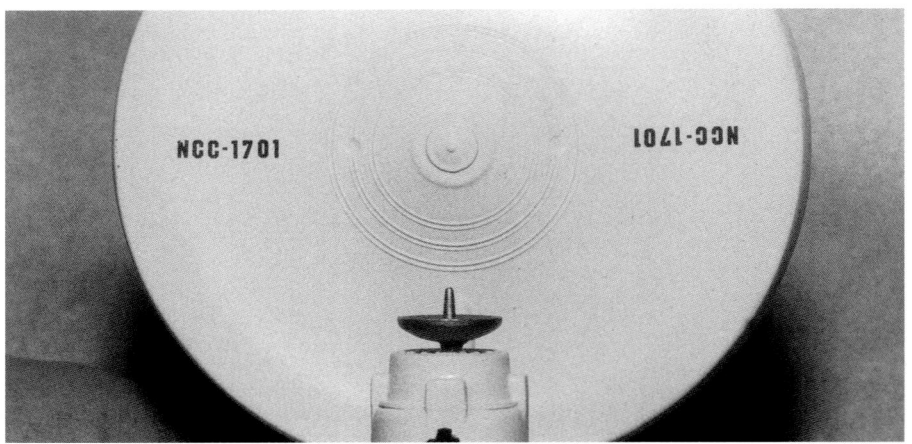

Fig. 1-17. Use the smallest characters on the decal sheet to add the ship's registry number to the underside of the saucer.

2
USS Enterprise: NCC-1701-E

The USS *Enterprise*, like the phoenix, will always arise from the ashes of destruction. After the -D was destroyed in *Star Trek: Generations,* the only question was what form the new *Enterprise* would take.

The *Enterprise-E* is a *Sovereign*-class starship. Dimensions are a little fuzzy, but it is longer than the *Enterprise-D* and not quite as wide. It retains the classic Federation lines while displaying an evolution in appearance. This results in a combination of the blended primary and secondary hulls, as in *Voyager*, and the basic layout that goes all the way back to the original series.

The kit of this incarnation is not overly complex to assemble. It is essentially the same as any of the previous kits. Your main effort will be directed towards the paint scheme. The "big screen" has allowed the special effects team to go to town on a complicated paint scheme, and so the paint scheme has become more complex as new ships are released. Be prepared to spend half of your time on this model in masking and painting.

Basic assembly

The *Enterprise-E* consists of 28 gray parts and 10 clear parts. This kit has some of the best fit found in the *Star Trek* line.

I immediately deviated from the basic instructions by cutting the socket from part 21. The peg for the base mounts into this, but I wanted to use a brass tube for a different style of base. I mounted the sleeve tube into the secondary hull with epoxy putty (fig. 2-1). Score the inside of the part with a knife or file to give the putty a textured surface to grip. Use two-part epoxy putty found at most hardware stores to anchor the tube.

Whether you use the kit base or one of your own, be sure to add weights to the engine nacelles as shown in Step 2 of the instructions. The model is too nose-heavy without them (fig. 2-2).

The kit has been engineered to eliminate seams that would force you to destroy surface detail when sanding the seams. This is most noticeable on the secondary hull. Instead of just two halves, there are separate plates for the rear and front sections on the bottom (parts 9 and 20) and a top piece (part 25) for the secondary hull. A "stair step" is formed around the joint once it is in place. Fill and sand only a portion of this seam, back to about the aft frame of the second recessed panel. Be sure to fill the forward portion of the joint, as this is visible once you've attached the saucer (figs. 2-3 and 2-4).

Other detailed areas are applique parts that hide the seams, such as the ventral phaser/sensor array (part 26). Cut a small notch about 2mm deep in the back edge of the head of the sensor (fig. 2-5). This will allow the sensor to overlap the deflector plate properly and lie flat along the belly of the secondary hull (fig. 2-6).

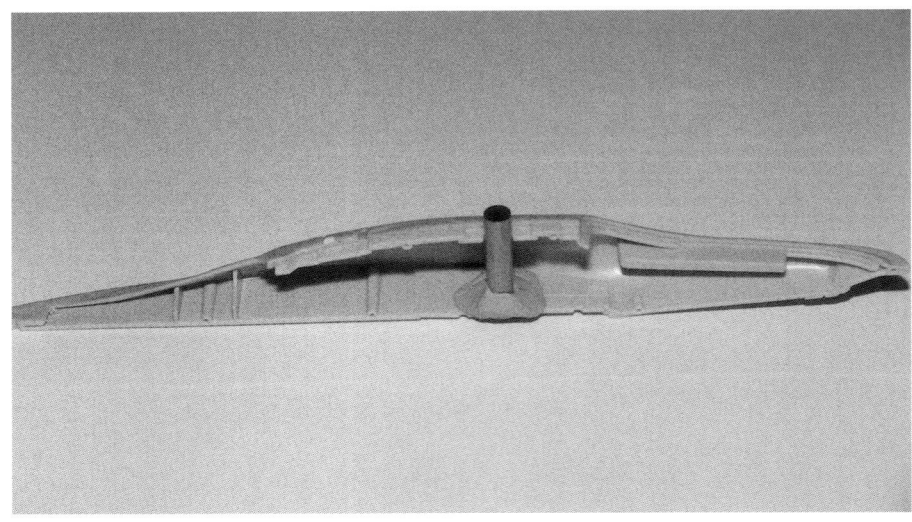

Fig. 2-1. Attach the mounting tube with epoxy putty and run it through the hole for the display stand.

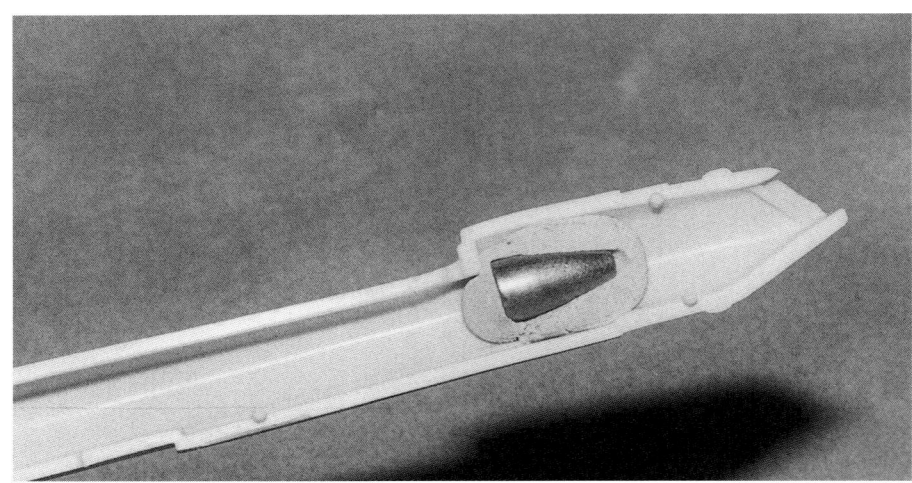

Fig. 2-2. Epoxy a small lead fishing weight weighing about 1 gram into the nacelle to balance the model.

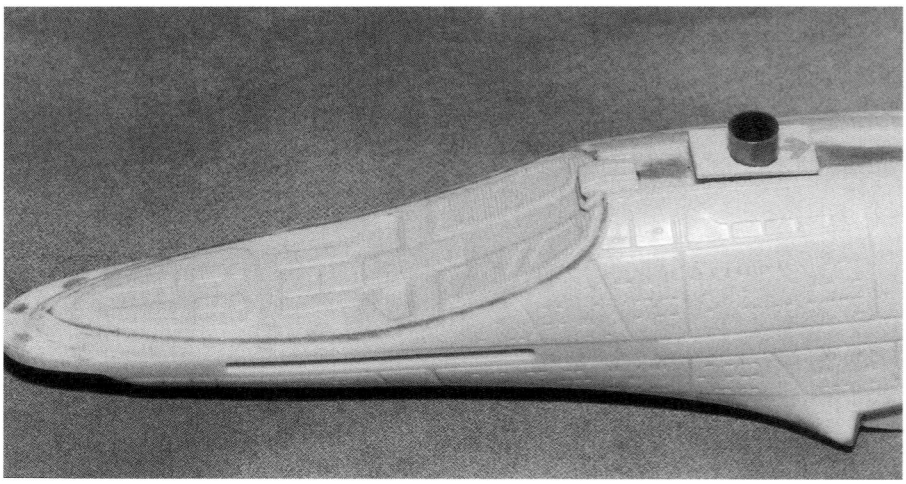

Fig. 2-3. Glue the aft plate in place.

Fig. 2-4. Use filler around the edges and sand smooth.

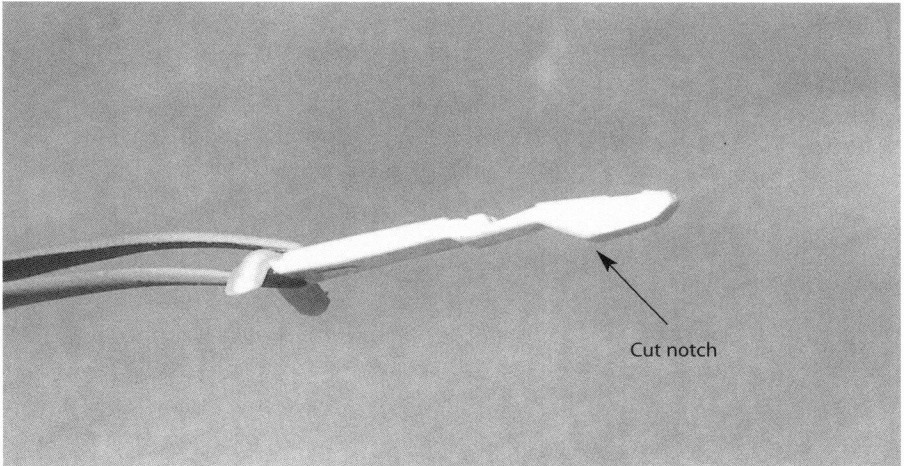

Fig. 2-5. Cut a notch in the back edge of the head of the array so it will fit over the edge of the deflector plate.

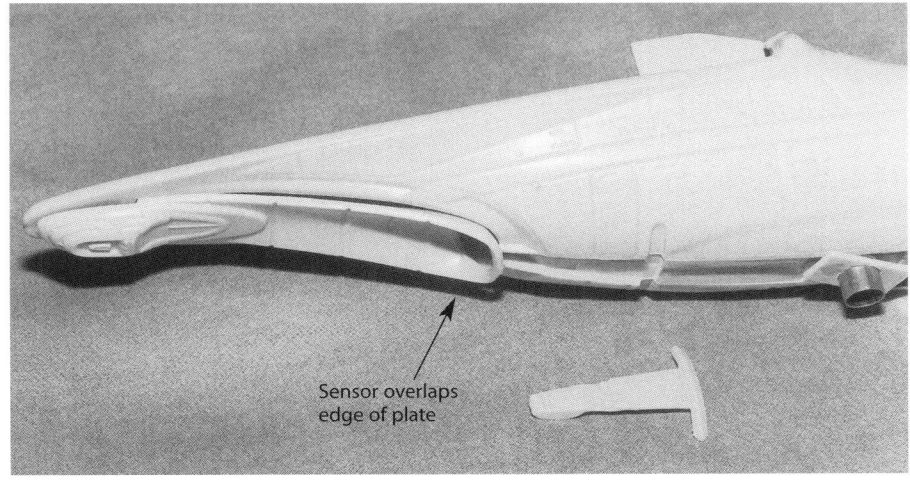

Fig. 2-6. Glue the phaser/sensor array on after the forward deflector plate.

Once the secondary hull has been glued together, except for parts 9 and 26, test-fit the forward deflector plate (part 9). It is slightly pear-shaped, so it will hit the sides of the opening. Sand material from the back side until the plate will slide into the hole (fig. 2-7). Trim from the plate rather than trying to resize the hole. Set the deflector plate aside until the deflector dish has been painted and attached.

Check for sink marks around the stern of the secondary hull near the shuttle bay and underneath where the torpedo bay is located. Fill and sand these areas as needed (fig. 2-8).

The clear parts provide other challenges. Paint the bussard collectors and the deflector with translucent acrylic paints (fig. 2-9). Both Tamiya and Gunze Sangyo sell a line of paints for this purpose. A brush can be used to apply the paint to the insides of the parts, minimizing brush marks. A thick coat will achieve a rich tone, but be sure to allow several days for the thicker coats of paint to cure fully.

Paint the bussard collectors and the nacelles separately before assembling them, then mask around the collectors for touch-up painting once you've filled and sanded the seams.

The main deflector painting instructions are incorrect. While previous variants of the *Enterprise* had blue deflector dishes, the -E's is yellow. For the main deflector dish "glow," I mounted a small, round craft mirror behind the dish. Unlike the bussard collectors, I could use the mirror here because there was more room to work in and the dish isn't an odd shape (fig. 2-10). I used the same technique on the impulse engines on the rear of the saucer (fig. 2-11).

Use this approach whenever possible because a mirror is more reflective than foil.

To achieve the glow effect for the bussard collectors and to blank off the view straight into the nacelles, add a shim of plastic behind the collectors (figs. 2-12 and 2-13). Painting the shim chrome silver will work; however, I recommend either finding a hobby shop that carries chrome Bare Metal Foil (used for the chrome on car models) or buying some metal foil adhesive, like Microscale Bare Metal Foil Adhesive, and applying household aluminum foil. Follow the directions and apply the foil to the shim, being sure to make the foil as smooth as possible.

Regardless of the approach you prefer, using a highly reflective surface will make any light entering the clear parts bounce back and provide the glow effect. This is a way to achieve some of the lighting effects without having to electrify the model.

The worst fit is in the aft ends of the engine nacelles. These plates don't fit very well, and getting them filled and sanded without damaging the details is difficult. Use tape or frisket paper to create a barrier when applying filler to prevent it from straying into the details. Not only do you need to fill and sand the seams, but the rib detailing starts on top of the engine and runs through the plate to the underside. Be careful to align the plate as best you can when gluing it into place so you can sand the rib to make it look like an unbroken feature (fig. 2-14).

On top of the nacelles, part 17 fits between the halves as a kind of fin. There is an excessive gap once it is mounted, so use strip styrene to fill the area.

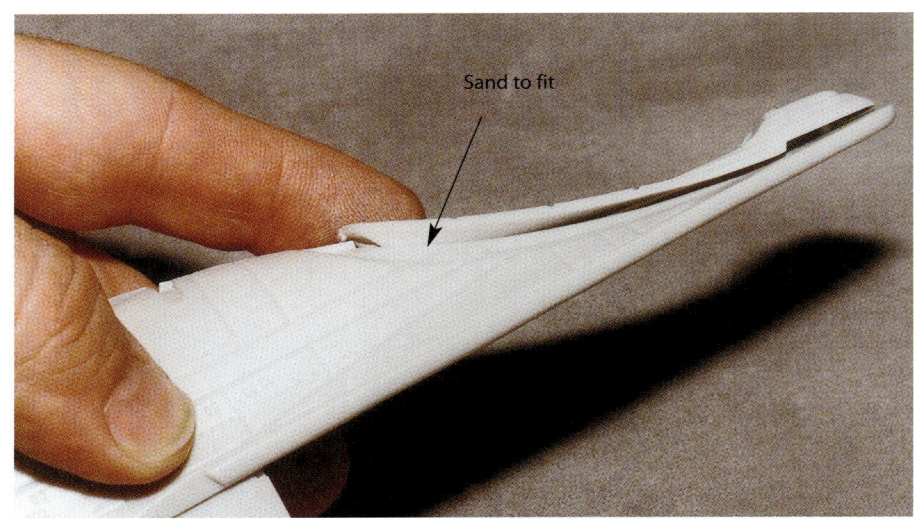

Fig. 2-7. You'll need to remove a significant amount of plastic from the underside of the deflector plate for it to fit into the hole in the secondary hull.

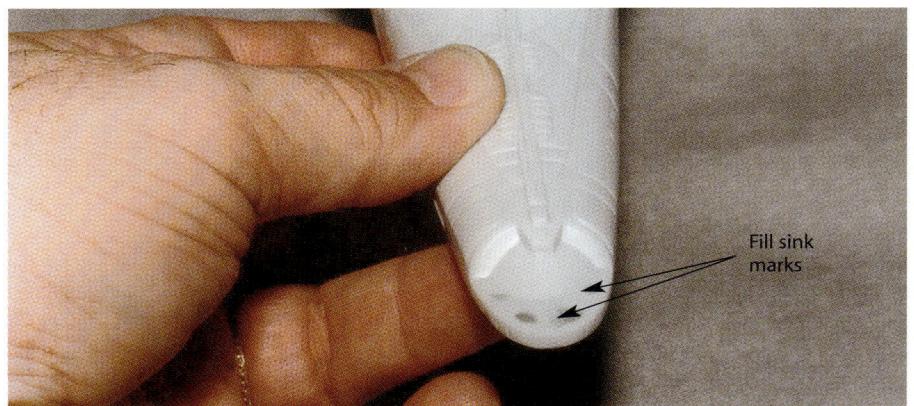

Fig. 2-8. There are more sink marks to be filled on the aft end of the secondary hull (shown before the shuttlecraft bay doors have been attached).

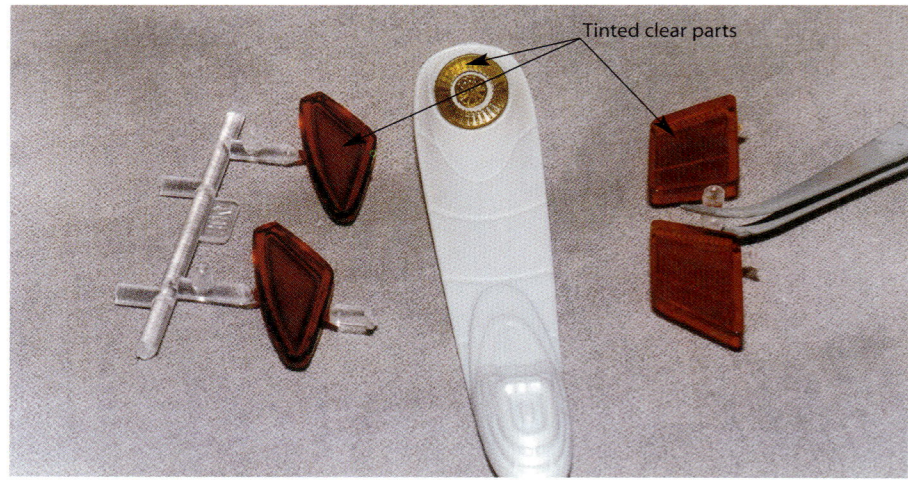

Fig. 2-9. Tint the clear parts using clear acrylic paints.

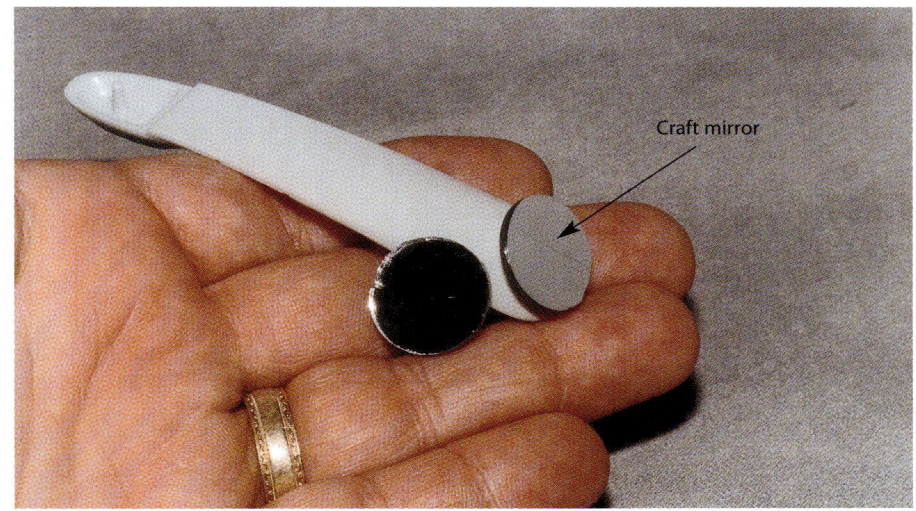

Fig. 2-10. By mounting a small mirror behind the main deflector dish, you can create the appearance of internal lighting.

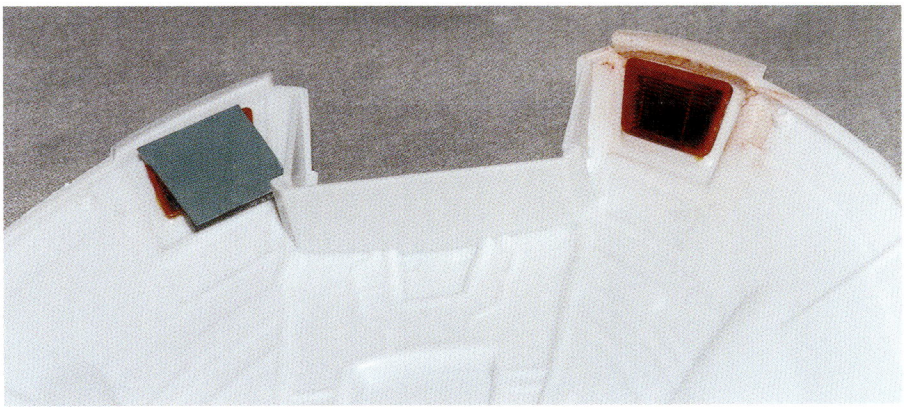

Fig. 2-11. Use the same technique on the impulse engines.

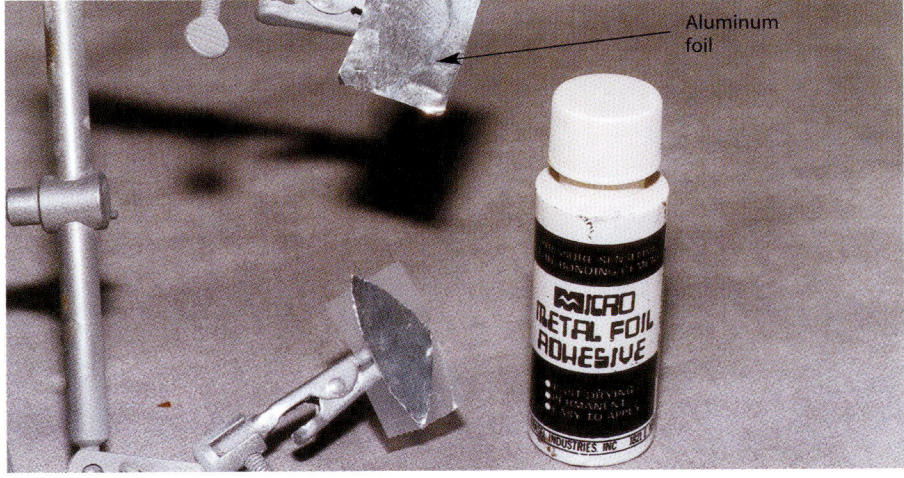

Fig. 2-12. Use metal foil adhesive to glue household aluminum foil (shiny side out) to shims sized to fit behind the bussard collectors.

You'll have to do quite a bit of filling and sanding where the nacelles are mounted onto the pylons. All of the gaps need filling, and the outboard edge of the pylon must be built up and blended to form an unbroken line with the area on the bottom of the nacelle (fig. 2-15).

Make the engines and pylons a subassembly and don't attach it until the model has been completely painted. This makes the masking and painting, especially of the secondary hull, easier. The joint for the pylon mount is a good one and won't require much touch-up. You may need to thin the mounting tab so it slides easily into the slot.

Painting

This kit has a more basic paint scheme than some of the other kits. Only four shades of gray make up the primary ship colors. In the ModelMaster line, Camouflage Gray FS36622 and Light Gull Gray FS36440 are recommended to make up the darker and lighter shades found on 98 percent of the model. Some details are the straight color, but for the overall model, both should be lightened by adding 10 percent white.

For the lightest shade, I felt that this was still too dark. I increased the ratio of white to 50 percent to get the base color closer to an off-white. For the darkest shade, the opposite was true. The paint wasn't dark enough. Shots from the movie give you an impression of a very dark gray with significant contrast to the lighter color. To get the effect I wanted, I substituted Gunship Gray, FS36118. Only the Light Gull Gray was used exactly as recommended in the instructions. Again, this is a matter of taste and interpretation, so experiment with the colors and find what you like.

Masking this model is an adventure. The lack of colors is made up for by the complexity of the patterns. I found that a combination of methods was required. To simplify masking around the curves and, especially, the narrow areas between the detail plates, use drafting tape. For this model, 1/32", 1/16", and 3/32" tape were appropriate; the 1/32" tape was perfect for the pinstripe areas. For larger masks, use the drafting tape as a starter around the curve; then use less-flexible tapes to form the bulk of the mask. If the area to be masked is large, but the surface is too irregular to use tape, substitute a water-based masking agent like Micro Mask. In all cases, start with the drafting tape to get a hard edge on the paint.

When applying multiple shades of paint, work from the lightest colors to the darkest. Dark colors cover better than light colors, so they should be applied last.

Spray the entire model with lightened Camouflage Gray paint. Whenever possible, apply no more than two coats a day and allow at least one day exclusively for drying to avoid marring soft paint. Mask off the areas to remain Camouflage Gray and then apply the Light Gull Gray (fig. 2-16). Repeat the process by adding more masking and then add the Gunship Gray as the last color (fig. 2-17). Remove the masking and touch up the edges of the panels as needed.

You'll have to carefully mask all of the clear parts, since removing their mask is the very last step in the painting process. A sloppy job will result in touch-up painting that will be difficult to blend with the overall paint job.

Part number 308 is a curved, clear piece that mounts in a groove along the back edge of the saucer halves. I have been unable to find

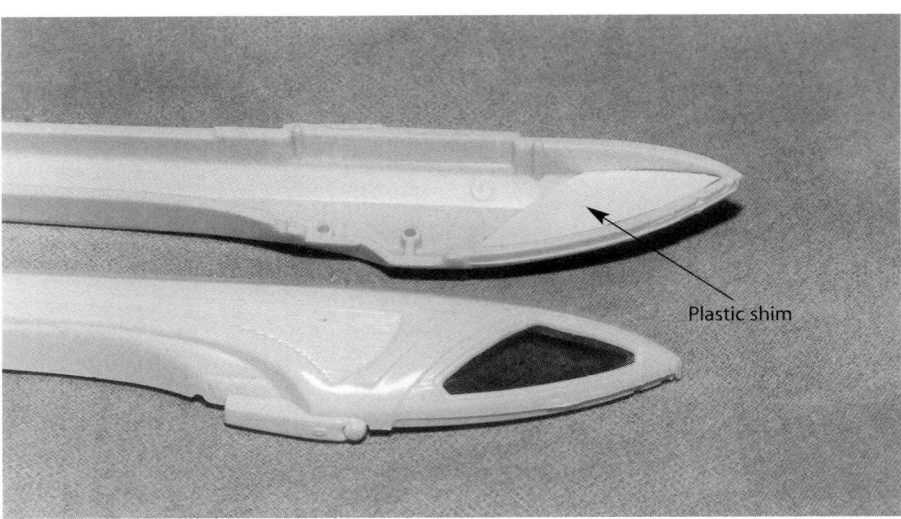

Fig. 2-13. Because of the odd shape, you'll have to use shims instead of mirrors in the engine nacelles.

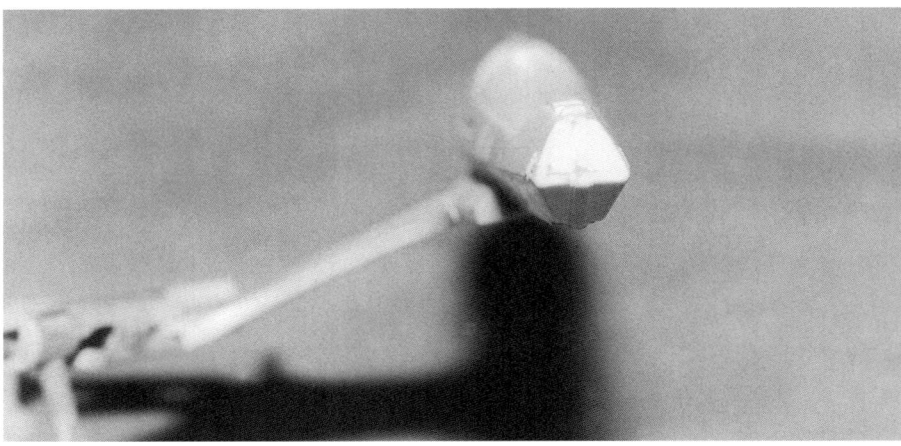

Fig. 2-14. Filling around the end plates of the nacelles is difficult. Be careful not to damage the detail while you're sanding the seams and blending the ribbing.

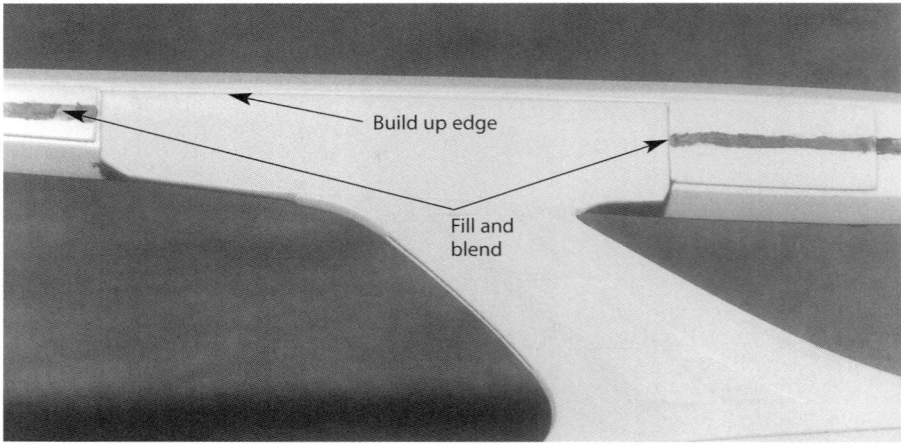

Fig. 2-15. Add shims of strip styrene to the edge, filling and sanding to create a continuous line.

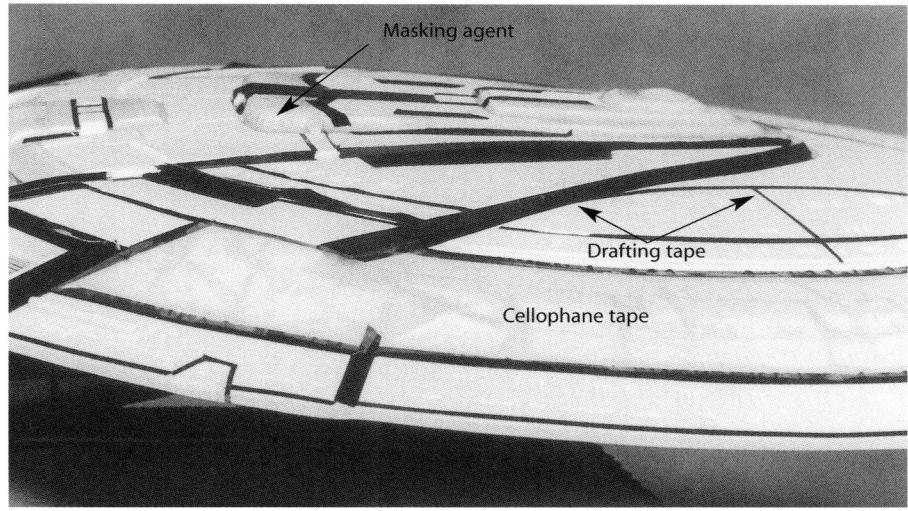

Fig. 2-16. You'll need a combination of masks to apply the paint scheme effectively.

Fig. 2-17. After applying the colors, remove the masks and touch up any areas where the mask leaked. Note the area where the impulse engines are located. The mask will remain until the very last step.

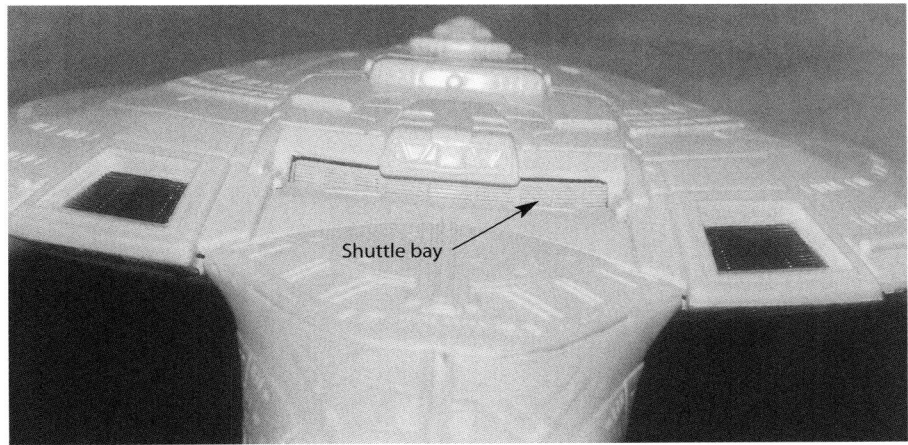

Fig. 2-18. Part 308 is a shuttle bay. Paint it the basic hull color with a pale blue for the grille areas.

out why it is clear or why the instructions say to paint it light blue. This is a shuttle bay, so it should be painted to match the hull, except that the oval details are a pale blue-gray along the lines of Duck Egg Blue or Russian MiG Light Blue (fig. 2-18). Don't worry about the gap along the sides or the bottom of the part once the saucer has been assembled.

Finally, paint the main engines a light blue that will allow you to maintain the glowing appearance. Here, I first masked and painted the center bar the Camouflage Gray. Then I masked it off and began to paint the clear area with a 50 percent thinned mixture of Pontiac Engine Blue Metallic, also from the ModelMaster line. The shade is right and the metallic finish in the paint adds to the glowing effect. Apply this in two or three thin coats of a 50/50 paint/thinner mix on the outside of the part until you are satisfied with the intensity. Spray a thicker coat on the inside to increase the opacity to the point that you can't see through the part. Remove the masking and gently touch up the areas as needed. Leave these off the model until you have applied the Dullcote.

The final painting step is to fill in the windows. Black and white are adequate, but you may want to throw in some yellow and blue-white to add some variety. The pattern is up to you, but you should keep windows that appear to be part of the same cabin the same color.

Use a .05mm water-based artist's pen to draw in the darkened windows. For the white, yellow, and blue-white, paint in the windows using a 50 percent thinned mixture of white paint applied with a 10/0 brush. You may have to apply several coats, since the thin mix won't cover well (fig. 2-19).

Decals

Prep the model by spraying it overall with one or two coats of clear gloss. Once this has dried (one to two days), begin applying the decals, working exclusively on either the top or the bottom at one time. Pay careful attention to the key numbers for the escape pods on the top of the saucer. Each has a unique shape; you need to be sure they are in the right locations.

Don't use setting solutions on these decals! The decals in *Star Trek* kits from the past ten years don't react properly to setting solutions.

Begin by applying a little water to the target area and then use a wet, soft brush to slide the decal off the paper onto the model (fig. 2-20). Use the water to float the decal while getting it into the final position. Once it is positioned, wring most of the water from the brush and then gently touch it near the decal to suck up excess water around the decal (fig. 2-21). You can also use the corner of a paper towel. This will keep the decal from floating out of position when you aren't looking.

If there is a significant surface contour under the decal, check on it every few minutes. Use the brush to push the decal around the detail, eliminating air bubbles. If necessary, add some water to the area to help re-float the decal, but be careful. Too many attempts will wash away too much of the adhesive and the decal won't stick.

After the decals have completely dried, put the model under some strong directional light and check it for glue residue and water spots. Clean them off the model before applying the flat coat. If you don't, they will leave an inconsistent finish, since they will be flatter than the clear-coated paint. Worse, over the years the glue will begin to age and turn brown. A starship with liver spots just doesn't work.

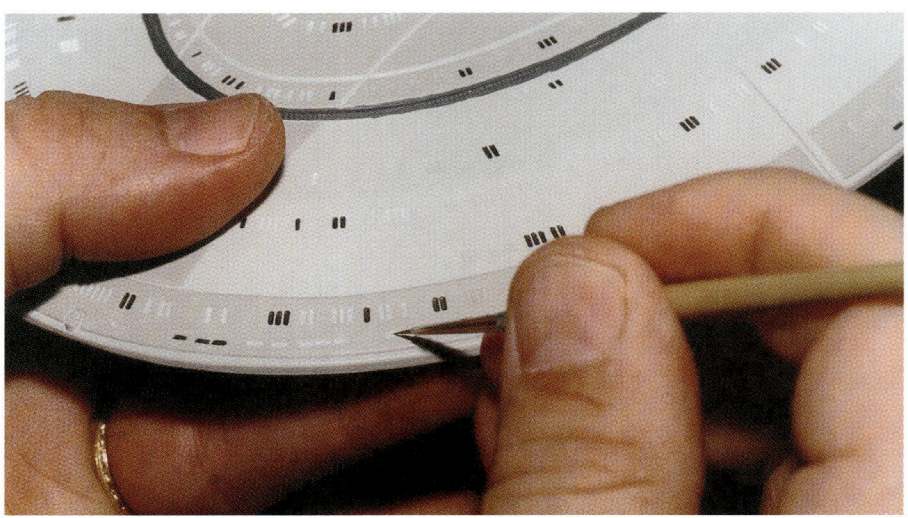

Fig. 2-19. It's tedious, but painting the windows is a must to make the model come alive.

Fig. 2-20. Use a large, soft brush to anchor one end of the decal as you slide it off the paper.

Fig. 2-21. Use the brush to drain the excess water from around the decal once it is in position.

3

The Klingon Bird of Prey Flight Display

The Klingon Bird of Prey was introduced in the movie *Star Trek III: The Search for Spock*. Since then, it has become the primary ship used by the Klingon Empire in both Next Gen and DS9. Small, heavily armed, and possessing a cloaking device, it is a very dangerous opponent.

The Bird of Prey (BoP) has been released in conjunction with each of the movies since *Star Trek III: The Search for Spock*. It has also been released in a Fun Pack containing paint and glue. A smaller version can be found in the Adversary Set.

The scale of the kit is listed on some of the boxes as 1/650. This would make the ship larger than the original *Enterprise*, yet manned by only 40-odd crewmen! A scale of 1/144 to 1/200 is more reasonable, especially when you consider scenes from *Star Trek VI: The Voyage Home* in which the crew is standing next to the ship. (It does make for interesting diorama possibilities, doesn't it?)

Basic assembly

Since a standard kit has been boxed with the flight display, both sets of "shoulder" grilles for articulating wings are in the kit. This gives you the option of modeling the BoP in a combat mode (wings dropped) or in a cruise/atmospheric mode (wings up). To use the flight display, you'll have to place the wings in the up position (fig. 3-1). If you put them in the dropped position, the opposite wing will hit the display disk and the model itself will be in a very awkward position. Be sure to refer to the numbers in the instructions to select the right set.

Even more critical is following the step sequence to the letter. While either wing can fit through the display disk, you cannot attach the disrupter cannon pod to that wing until after you slide it into the notch (fig. 3-2).

The "shoulders" form the attachment points for mating the wing to the fuselage. After assembly, there is a gap on the underside. Use strips of styrene to create a blank-off (fig. 3-3). Add the styrene in sections so you can follow the contours of the shoulder piece. Fill and sand the blank to eliminate any gaps and seams.

You'll have to blank off the notch for the standard base, since it won't be used for this version. Glue a piece of .040 strip styrene into the slot, then fill and sand smooth (fig. 3-4).

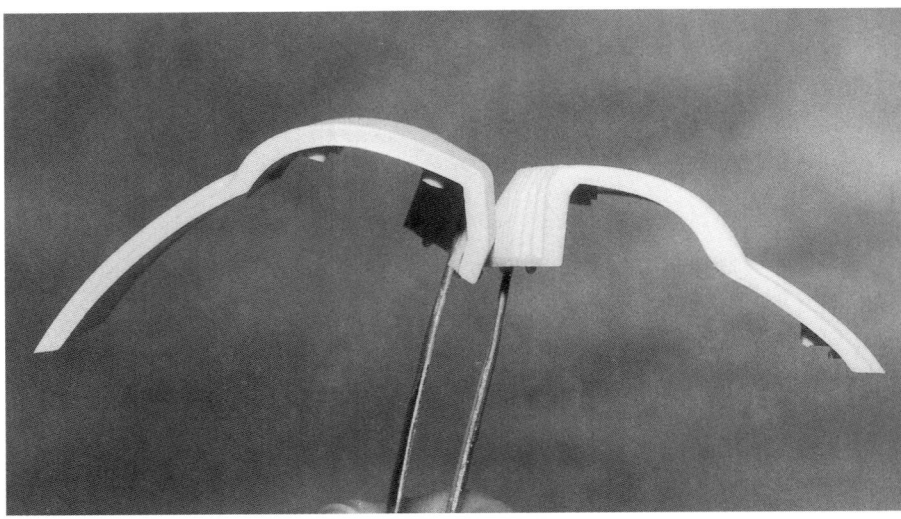

Fig. 3-1. The longer shoulder panels (on the left) are for the combat mode of the kit and can't be used for the flight display.

Fig. 3-2. The wing will fit into the slot in the acrylic disk at an angle. You can use either wing.

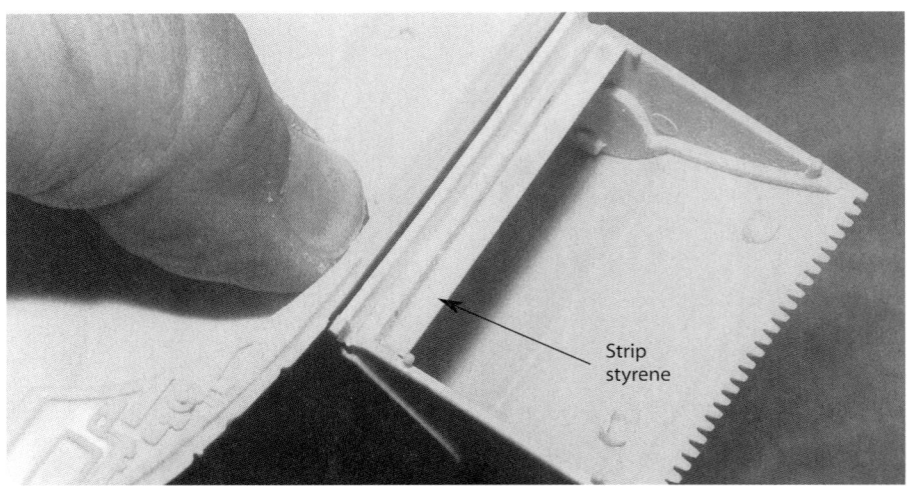

Fig. 3-3. Use strip styrene to fill the gap between the fuselage and the inner edge of the wing.

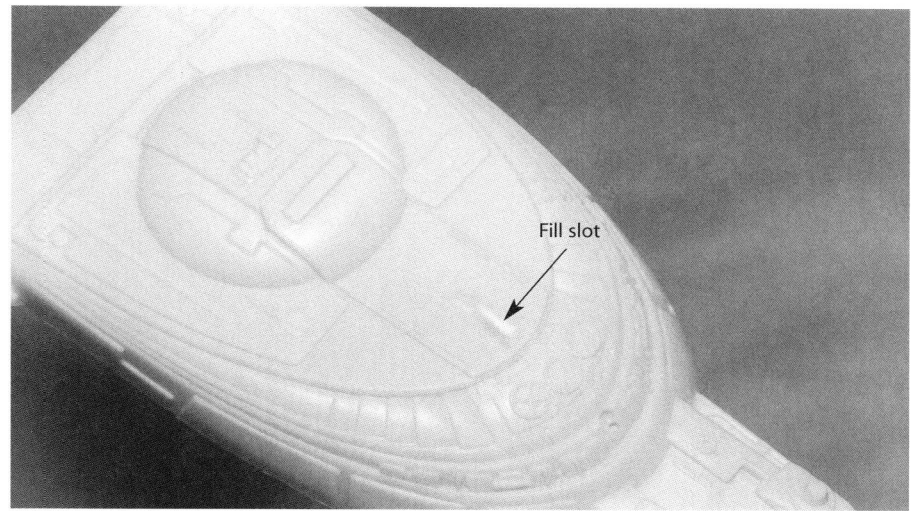

Fig. 3-4. Fill the slot for the standard kit base on the underside of the hull.

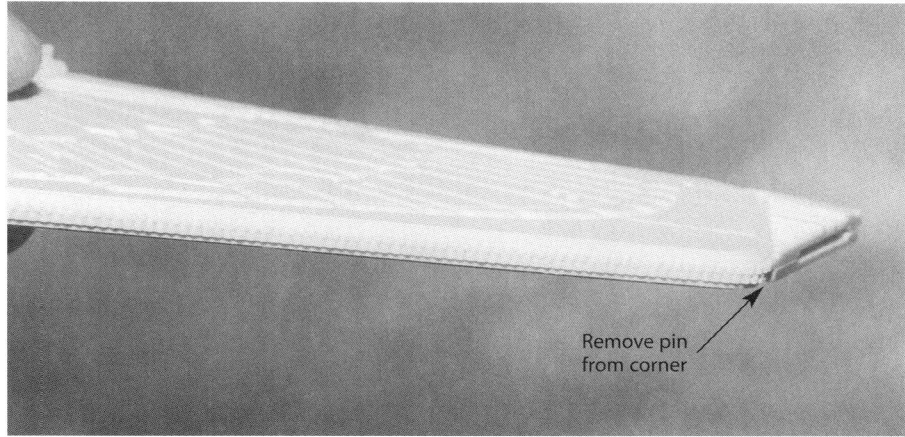

Fig. 3-5. Remove the alignment pin so the corrugation detail will line up.

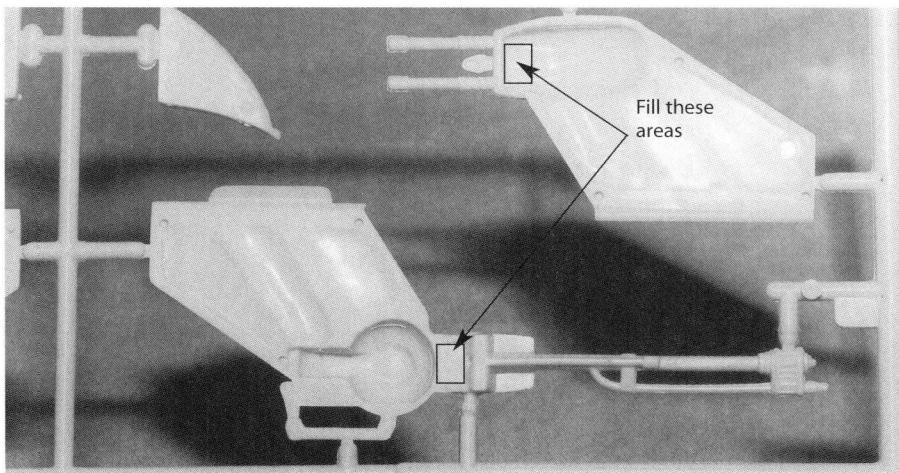

Fig. 3-6. Fill and sand the inside portion of the cannon halves.

The trailing edge of the wings has a corrugated effect. When test-fitting the wings, you will find that the pattern on the top resists lining up with the pattern on the bottom (fig. 3-5). Cut off the alignment pin at the outboard rear corner of the wing so you can force the corrugation to align, and glue the halves together.

The insides of parts 15, 16, 17, and 18 (fig. 3-6) for the disrupter cannon pods also need filling. If you don't fill them, a hollow area will be visible all the way through the gun pod after the halves are glued together (fig. 3-7). Use plastic to form most of the mass. Using filler alone risks softening the plastic and having it distort the plastic as it dries. The more filler you use, the more shrinkage you will have, requiring you to add additional coats.

Drill out the barrels for the disrupters to a depth of 1 to 2 millimeters. For the large guns, use a 60-gauge drill bit in a hand drill. The smaller guns need a 70-gauge bit (fig. 3-8). After painting the guns, add a drop of India ink or black paint in the depression to complete the depth effect.

Sometimes seams are impossible to fill with sanding fillers because of the extreme angles. If the gap is small, try painting in some white glue (fig. 3-9). One or two applications will usually do the trick.

Painting

The painting chart lists several colors, but it tells only part of the tale. Colors "A" and "B" are green and gray, respectively. Yet as you look at the instructions, you will find that several shades of each color are associated with each "A" and "B." The instructions identify the colors by their FS numbers (see Appendix 1) whenever they tell you to apply paint. Using the ModelMaster names, the colors are:

Gray
FS 36375 Lt. Ghost Gray
FS 36118 Gunship Gray

Green
FS 34227 Pale Green
FS 34159 SAC Bomber Green
1734 Zinc Chromate Green

This model requires you to paint in stages prior to assembly. Even when you're spraying a single color, there are too many odd angles. If you wait until the kit is assembled, you will risk "dead spots" or overspraying details that cannot be masked properly. Paint parts while still on the sprue whenever possible (fig. 3-10). This makes handling much simpler. Be very careful to turn the parts to all angles as you paint them to prevent any dead spots. Before proceeding to the next step in the assembly, take the painted parts to a good, strong light source and check them for missed spots. As you glue the painted parts and fill the seams, you'll have to do some touch-up painting (fig. 3-11), but that is an easy problem to deal with.

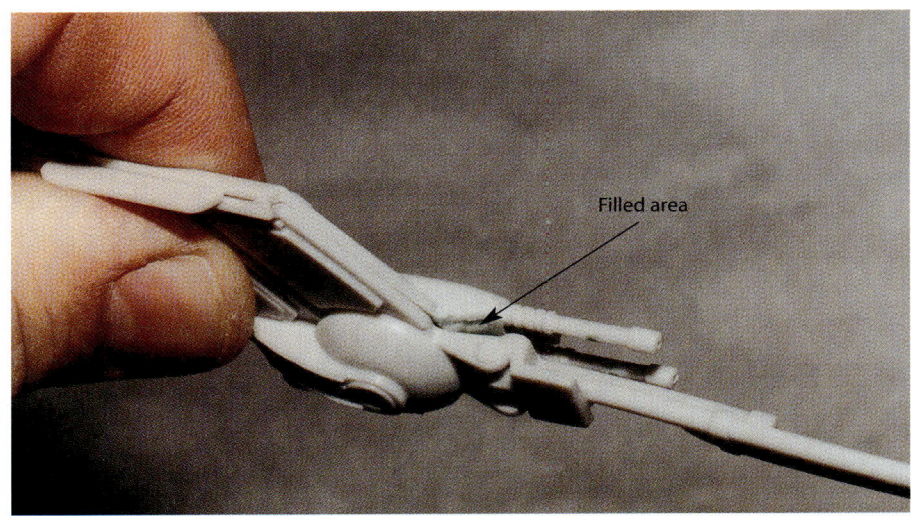

Fig. 3-7. The filled areas prevent seeing all the way into the gun pod.

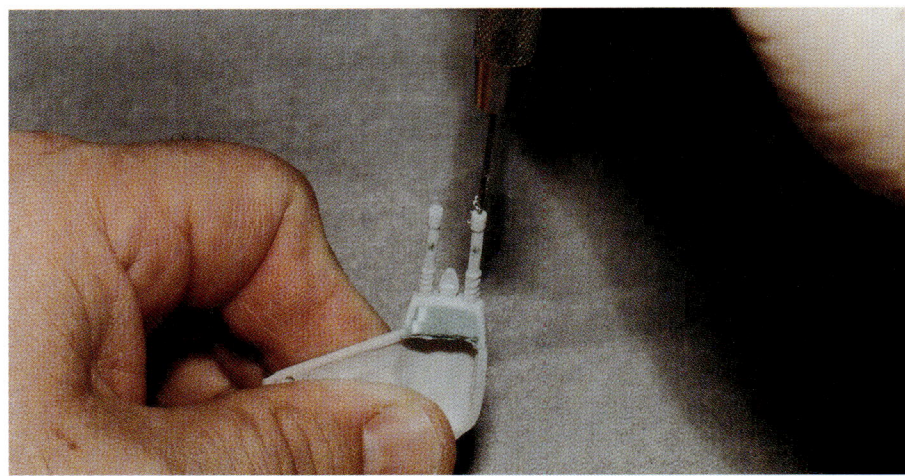

Fig. 3-8. Before gluing the gun pods together, drill out the gun barrels using a hand drill.

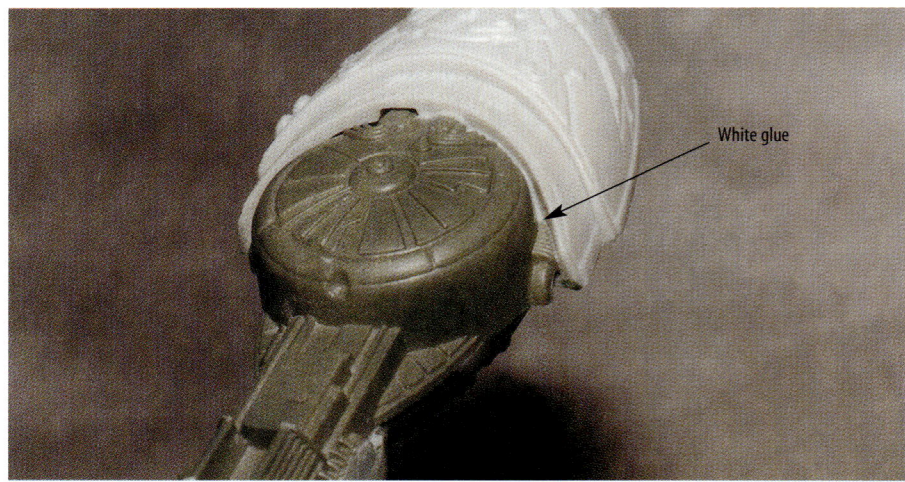

Fig. 3-9. In a tight area, you can fill an irregular gap with white glue instead of filler.

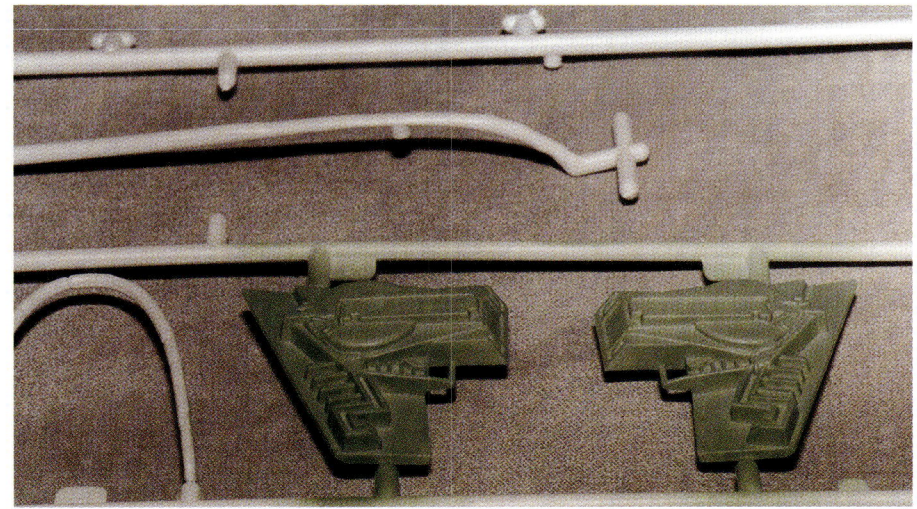

Fig. 3-10. When painting the parts, shoot from all angles to get paint into all of the nooks and crannies.

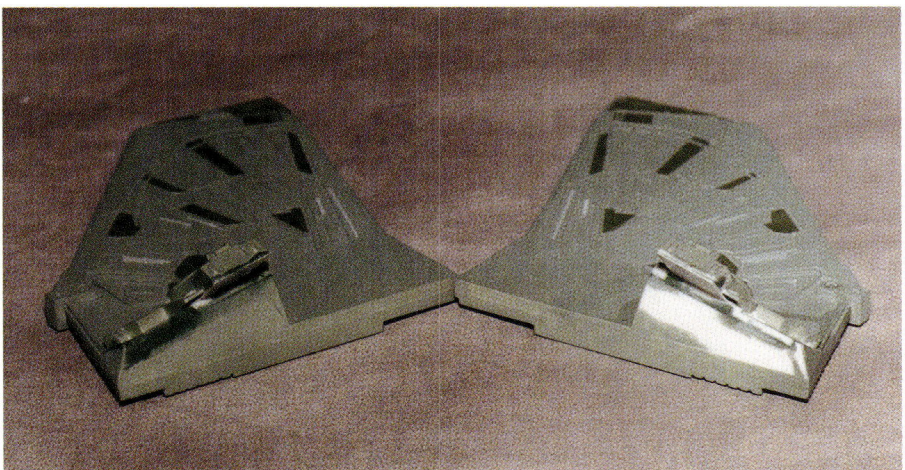

Fig. 3-11. Fill and sand the seam between the wing and the detail plate.

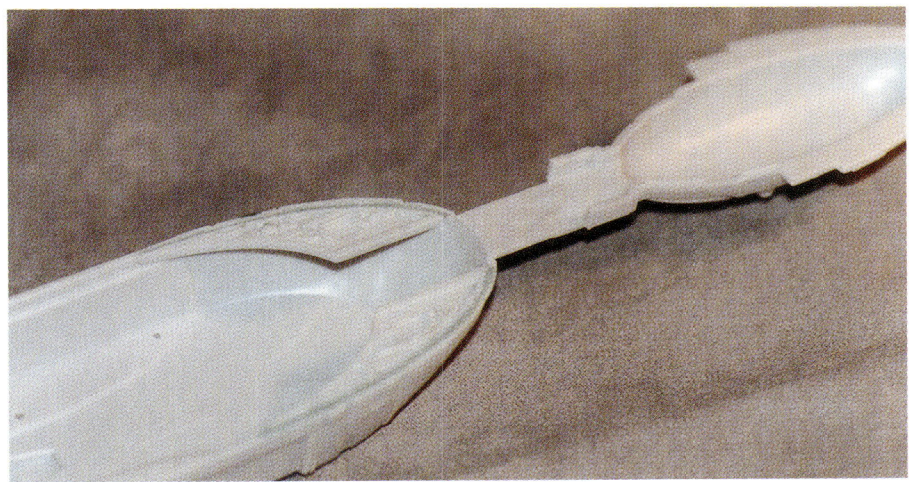

Fig. 3-12. Attach parts 9 and 10 to the upper hull, fill the seams, and paint.

Parts 9 and 10 are painted steel, but they are located right up against other parts painted green and are in very tight quarters. Attach parts 9 and 10 to the upper hull (part 1), fill and sand the joint (fig. 3-12), then paint the parts steel. Attach the pre-painted spacer ribs (parts 3 and 4), and then cover the steel paint with a masking agent like Micro Mask to protect it from the paint being applied to other parts (fig. 3-13).

You should also use a sealer for metal painted parts (like Testors Metalizer Sealer) to protect the finish of parts that will receive a lot of handling after they have been painted. The "shoulders" are painted separately, but still must be attached to the wings and the fuselage. In doing this, you risk rubbing the metal paint off the parts.

After assembling the upper and lower hull and the spacer ribs, you will find a rather ugly joint along the sides of the neck where all three parts meet. Be sure to fill and sand this area before painting (fig. 3-14).

Farings 19 and 20 have prominent ejector pin marks on the underside, and there are three mounting pins on each faring to attach them to the upper hull. The ejector marks and the bulge surrounding each pin on the inside of the faring are very noticeable during a test fit, but they are not visible after you mount the shoulders.

Attach parts 21 and 22 to the farings and paint them steel. Mask these parts before applying the FS34227 green to both the inside and outside of the farings.

The Bird of Prey emblem is painted with ModelMaster Sunburst on the underside of the wings. To protect the separations between each feather, apply a masking agent along the seams (fig. 3-15). Protect the larger areas around the feathers with tape (fig. 3-16).

Since Klingon ships have a tendency to be "weatherbeaten," airbrushing the insignia is essential. Apply lighter coats and stop while the green undercoat still shows through. Irregularities and mottling of the finish are very appropriate. Keep the feel of the logo consistent with the overall weathering of the model.

Peel the mask away from the wing (fig. 3-17). If some of the sections prove too difficult to remove because the mask is too fragile, all is not lost. After the paint has dried *completely* (especially if you are using water-based paint instead of enamels or lacquers), get a bowl large enough to hold the wing and fill it with cold water. Soak the wing in the water for about an hour. The mask will reliquefy enough to lose its adhesion. You can then remove it with a stiff brush, scrubbing just hard enough to remove the mask but not the paint.

The engine assembly is clear part 400. It is not obvious why this and the photon torpedo launcher were molded in clear

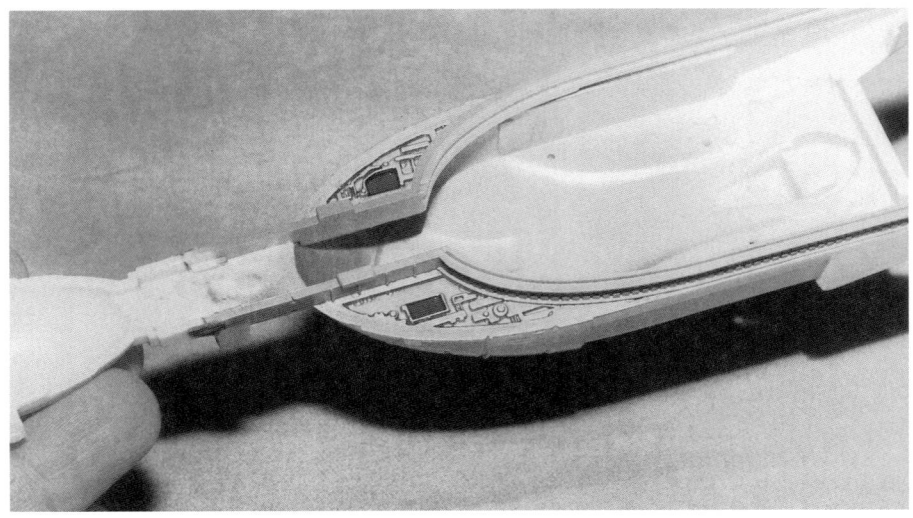

Fig. 3-13. Use a liquid mask to protect the steel painted panels from the other colors as you paint the rest of the model.

Fig. 3-14. There is a bad fit where the upper and lower hull halves and the spacer ribs come together on the neck of the model.

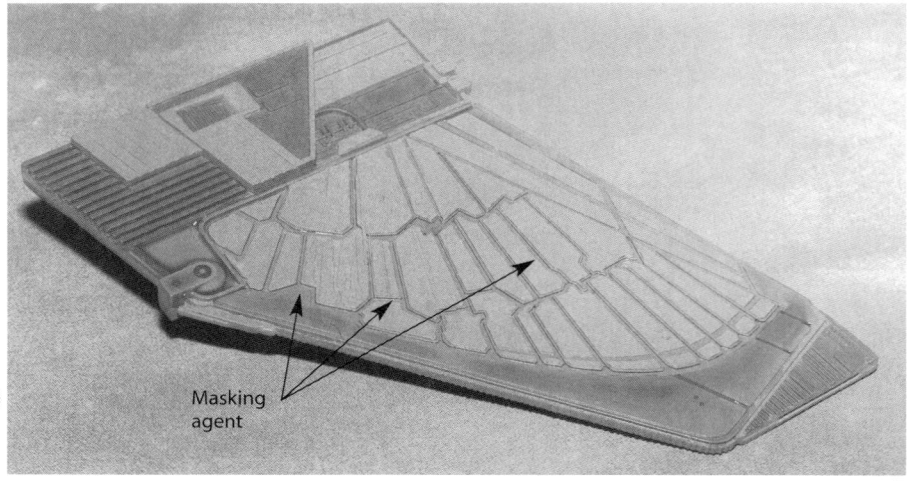

Fig. 3-15. Mask the separation lines before painting the feathers for the bird of prey emblem.

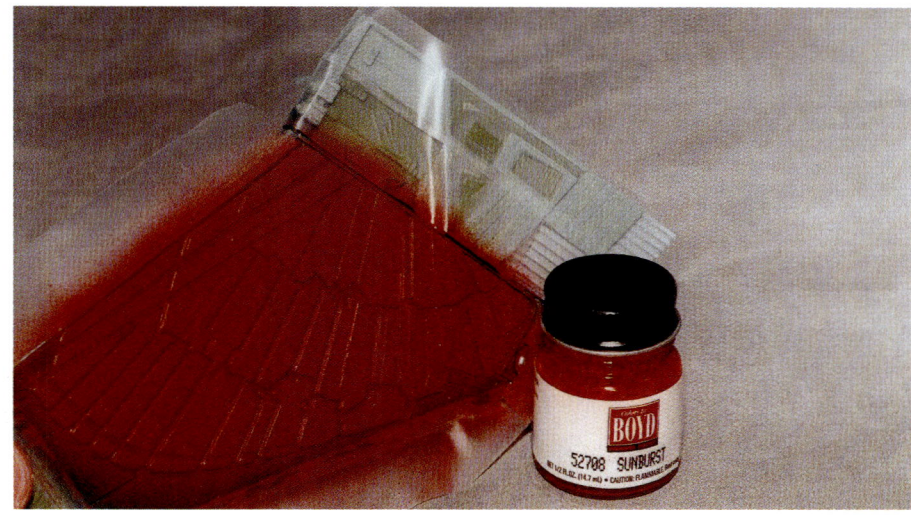

Fig. 3-16. After protecting the outside edges of the wing, paint the feathers.

Fig. 3-17. Peel off the mask. Touch up the pattern if needed.

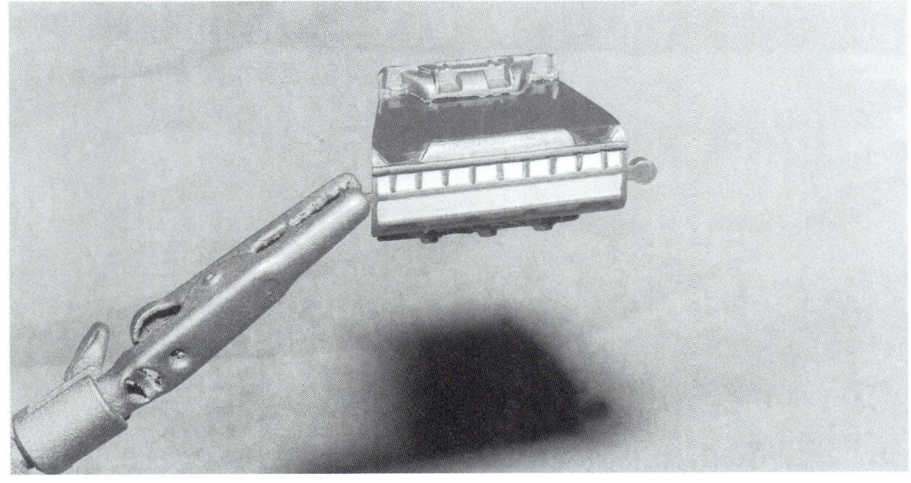

Fig. 3-18. The painted engine

plastic. Perhaps at some point the company contemplated a lighted kit. Take advantage of it by back-painting the red areas to create a depth effect. Next, mask the areas where the red is to show through, and spray ModelMaster Stainless Steel paint on the exterior. Apply the yellow and white paints with a small brush. The yellow I used was a fluorescent artist's tempera to simulate a glowing effect (fig. 3-18).

Decals

There are only four decals for this kit, but they will take quite a bit of time to apply. Even though this is a fairly contemporary kit, the decals are old-style—thick and hard to apply properly.

A smooth, glossy surface under the decals is critical. Because the decals are so stiff, any irregularities under them will invite silvering.

Making matters worse, the insignias on the gun pod pylons must conform to an extremely irregular surface (fig. 3-19). While a setting solution on most of the newer *Star Trek* kits is not recommended, here it is a must, and in large quantities.

Once you have applied the insignia, don't forget about it. Push it into the corners and recesses with a damp brush. Reapply the setting solution to soften the decal until it follows the shape of the part. If all else fails, use a razor blade to carefully cut the decal at the point at which it should make a sharp corner to allow it to lie flat.

The wash

For this model I used an India ink wash to highlight the surface detail.

Create the wash by filling a small container with denatured alcohol. A film canister is a convenient size and will provide more than enough wash for this project. Add three to seven drops of black India ink. Fill a second container with plain denatured alcohol.

Select an area to work on, such as a wing, where there is a sharp corner or ridge to separate the area from the rest of the model. With a large brush, lay down a coat of the clear alcohol (fig. 3-20). This will allow the wash to flow completely and evenly rather than starting to dry too soon. Then, using a different brush, apply the wash from nose to tail or leading edge to trailing edge—whatever is consistent with the normal flow over the model (fig. 3-21). Let the ink puddle and flow where it will and wait until it has dried. If the intensity of the weathering is satisfactory, proceed to another area until the model is completely weathered. To weather areas in tight quarters, use a smaller brush (fig. 3-22).

Using this technique, you will find that once the wash has completely dried, the area treated has become chalky (fig. 3-23). This is the result of an interaction of the alcohol and the lacquer clearcoat used to prep the decals. Don't panic! You still have a final overcoat of flat to apply. This will cause the chalky areas to disappear, leaving the darkened areas visible.

If the wash is too dark, you can

Fig. 3-19. The decals need a lot of work to get them to fit over the irregular surface.

Fig. 3-20. Apply a heavy wash of alcohol to the model to prep for the weathering wash.

Fig. 3-21. Apply the wash in the "airflow" direction of the model.

Fig. 3-22. Also apply a wash to the recessed areas on the shoulders.

Fig. 3-23. Chalking is normal. It will disappear when you apply the final overcoat.

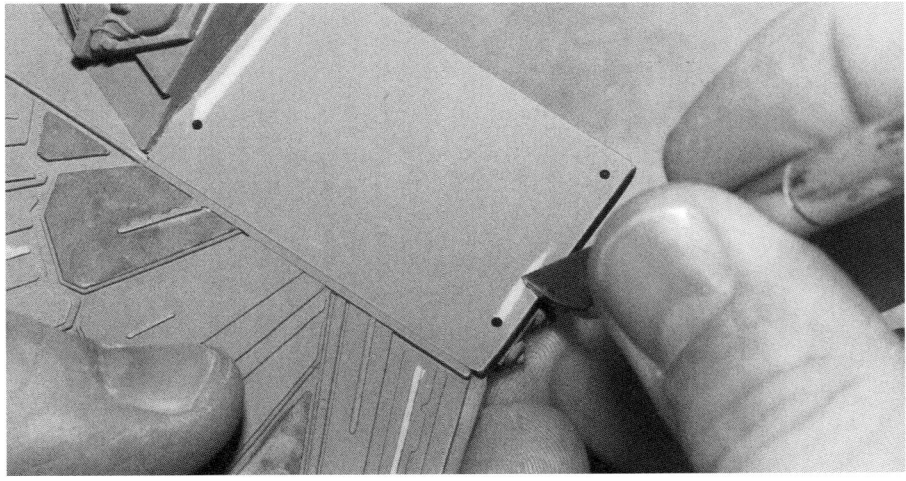

Fig. 3-24. Scrape the paint off the surfaces to be glued.

quickly apply more clear alcohol to reliquefy some of the pigment. Squeeze most of the alcohol out of the brush and pick up the excess pigment by touching the brush to the area and letting the capillary action suck it up.

To change the intensity of the wash, pour out some of it and replace it with straight alcohol.

Final assembly

Join the wings to the shoulders first so you can apply the glue to the inside of the joint. This needs a clean plastic-to-plastic surface. Apply a little extra glue, since this is a primary load-bearing joint. Scrape the paint from the wing where it will join the shoulder (fig. 3-24). Working from the inside, apply glue to the joint. Don't be stingy! You can be aggressive because working from the inside reduces the chance of damaging the exterior details or paint.

Mount the wing-shoulder assembly to the top of the hull. Again, apply extra glue to make the joint solid. The area under the farings is the best spot for this, since this area is not visible once you attach the farings.

Attaching the farings is now the trickiest part, since you'll have to deal with that seam after attaching the other parts. Sand and test-fit each faring before trying to attach the wings to achieve the best possible fit.

To keep the model from being over-balanced when placed in the flight display, epoxy several ounces of lead weights into the base.

4
Deep Space Nine

Deep Space Nine is the former Terak Nor station constructed by the Cardassians to orbit Bajor during their occupation of the planet. Once the peace treaty was signed, it was taken over by the Bajorans and placed under the administrative control of the Federation. When a wormhole was discovered, the station was moved out of orbit and positioned near the mouth of the wormhole to control the traffic moving through it.

Since it is Cardassian in origin, the station exhibits an architecture very different from anything previously seen in the *Star Trek* universe. While visually interesting, it is really a curse. This is a poorly fitting kit. To make matters worse, most of the fit problems are in areas that are extremely difficult to repair.

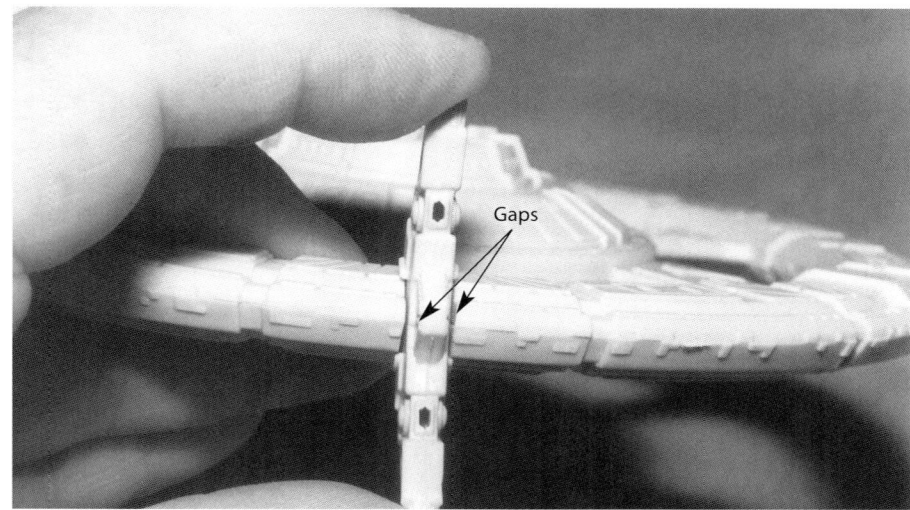

Fig. 4-1. You'll have to fill the gaps for the sails. Be sure the remaining slot allows the sail to stand vertically.

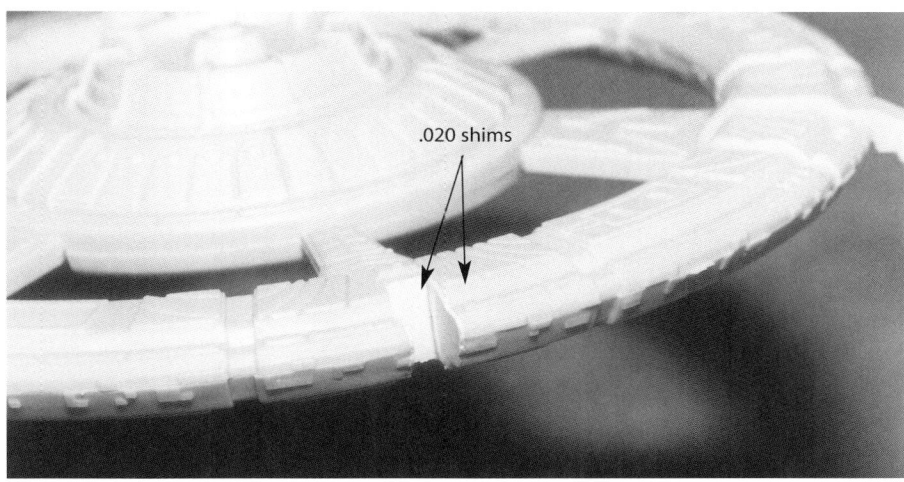

Fig. 4-2. After adding the shims, glue the sail into place.

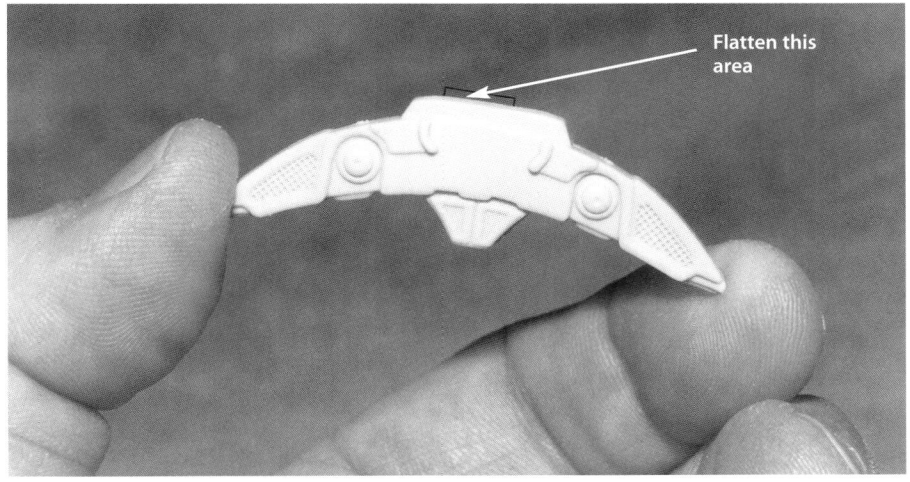

Fig. 4-3. To get the sail to fit flush against the slot, flatten the inside curve.

To describe the construction process, I have to provide a glossary of some of the station components. The central hub contains the command center, the promenade, and the main reactor. Outside of that is the inner habitat ring. The upper and lower pylons are mounted to the outer docking ring. Several small and large docking ports are also found on the docking ring. Finally, the habitat ring contains the "horns" (called "sails" in the instructions) for the photon torpedo launchers, tractor beams, etc.

Basic assembly

The shape of the station required some unique kit engineering. The hub and habitat ring combine to form a top and bottom clamshell. The fit here is reasonably good, but the habitat ring may be slightly warped. It may not be so noticeable at first, but the droop becomes more obvious when the docking ring is attached, so test it on a flat surface and be prepared to force it into the proper shape as you assemble it.

The sails are probably the easiest pieces to clean up and don't fit too badly. When you try to mount them to the habitat ring, the first hint of the fit problems appears. There are substantial gaps between the sail and habitat ring, which you'll have to fill (fig. 4-1). The bumps, lumps, and other details make it hard to get to the seams for filling and sanding. In this case, the method I found that worked the best was to add .020 styrene to the inside of the notch (fig. 4-2). Sand it down until the sail fits snugly on both sides. To get a good fit on the rear of the sail, sand a flat spot on the curved section (fig. 4-3). With a little reshaping, it seals the gap nicely.

No mounting pins are provided, so pay close attention to the alignment of the sails as you glue them into place. Make sure that the points form a vertical line and that they don't tilt to the side. The final step is to add a small bump to the top and bottom of each sail. The bumps will become strobe lights once the model is painted. Use white glue or epoxy for this effect.

The docking ring is also a top and bottom clamshell. It comes in three sections—one for each spoke. The ends of the spokes fit into recesses on the underside of the habitat ring.

The seams for the spokes are easy to fill and sand but tend to split because of the flexing of the spokes as you handle the model. Reinforce each spoke with some rigid bars (fig. 4-4). At the very least, I would suggest using extra glue or super glue to join these seams.

The outside seams on the docking ring fit very well and require little sanding. The inner seams are atrocious. On the inside of each section of the docking ring are three square areas that are cargo bays associated with each small docking port. Two of them have a reasonable fit, but the one closest to the spoke has a part mismatch of almost $1/16$" (fig. 4-5). I found it impossible to repair the fit without totally removing all of the detail (fig. 4-6) and replacing it after filling the seams. I used a combination of .02 x .02, .01 x .04, and .01 x .06 strip styrene to replace the removed detail (fig. 4-7). This surgery will probably require you to do the same to all three, because matching the rework to the kit-provided detail is very difficult.

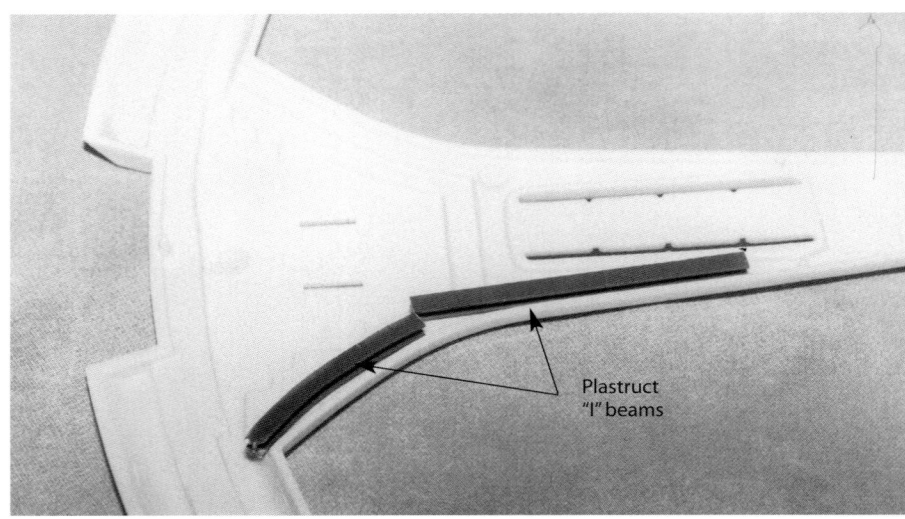

Fig. 4-4. Reinforce the spokes with epoxy putty or a spar to reduce the flexing.

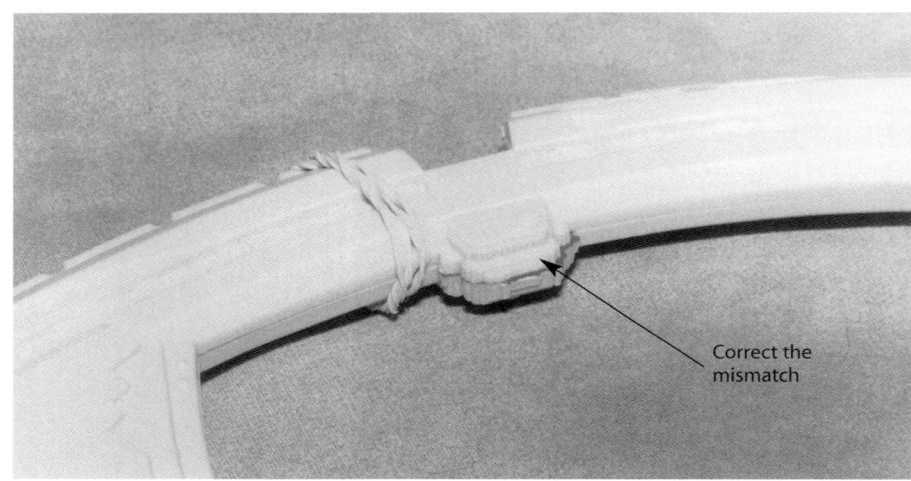

Fig. 4-5. There is a significant mismatch between the upper and lower parts of the docking ring.

Fig. 4-6. Remove the kit details and sand the bays.

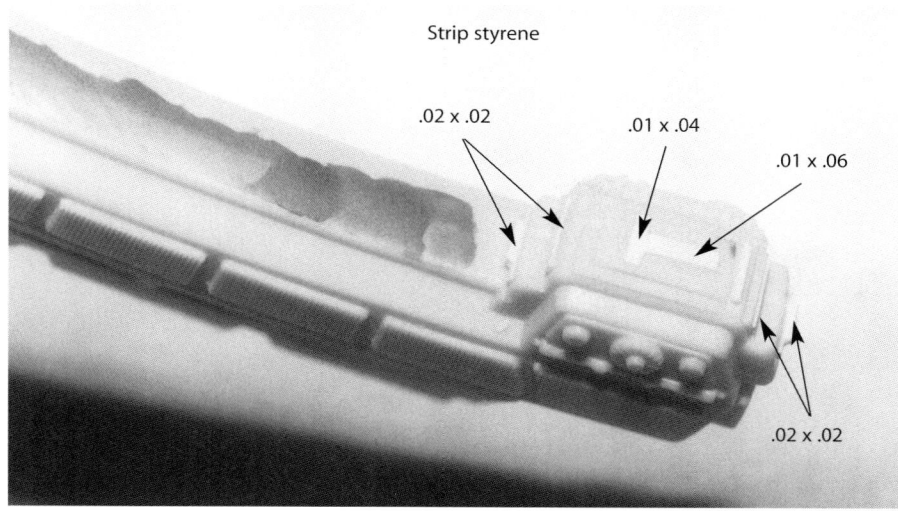

Fig. 4-7. Use strip styrene to replace the kit details.

Fig. 4-8. Place the habitat ring on a CD case when gluing the docking ring into place to get a consistent angle.

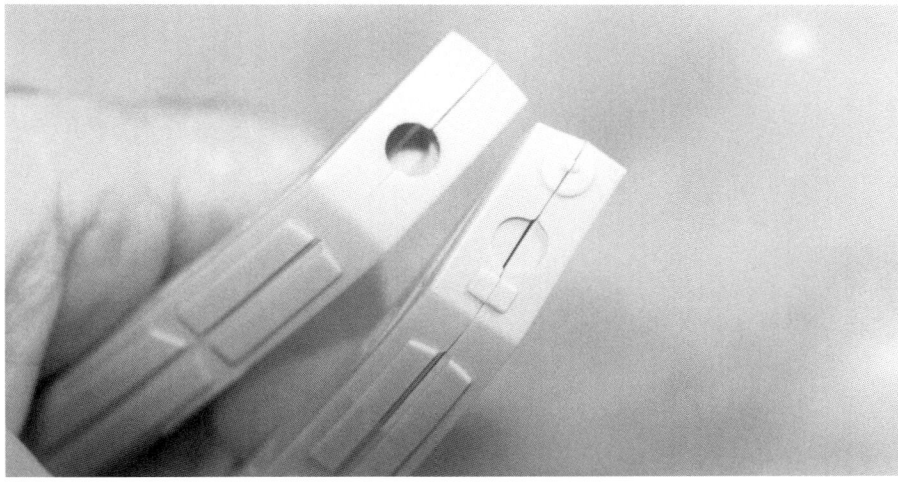

Fig. 4-9. The lower pylons have holes for the base, while the upper pylons have a circular depression.

After assembling the habitat ring, you will notice that it sits higher than the docking ring. This is accurate; just be sure that it is consistent after you attach all three spokes. As the instructions suggest, place the central core (before you attach the reactor section) on a raised prop while the spokes rest on a tabletop at the proper angle (fig. 4-8). Glue them together while they're in this position. Be prepared to spend time filling and sanding the seams between the spokes and where they join the habitat ring.

As you begin to work on the upper and lower docking pylons, note that the upper pylons have a circular depression, while the lower pylons have holes for the kit base (fig. 4-9). Regardless of how you will display the model, you'll have to fill the depressions in the upper pylons and sand them smooth (fig. 4-10).

If you choose to fill the holes on all of the pylons, be sure to mark them so you can tell the upper ones from the lower ones. The mounting tabs are generic, making them appear to be interchangeable, but you'll really mess up the fit if you try to put a pylon on the wrong side. The top of the docking ring has a shoulder that the bottom side doesn't have (fig. 4-11). You can't put the docking ring on upside down, but you can attach the pylons incorrectly.

The pylons come in a left and right half and also fit poorly. Hiding the seam along the outside curve is not very difficult, provided you use enough filler (figs. 4-12 and 4-13), but the inner seam is hard to sand because of the curvature of the part. Using anything having a straight edge means you have very little in contact with the part as you sand it. The sanding process takes several times longer as a result. On the inside of each pylon, consider using a riffler file (fig. 4-14). The curve of the file is similar to the curve of the pylon, giving you more surface contact. To prevent fouling the grooves, you'll find that wet-sanding and frequently washing out the residue work best.

While the outer seam is easy to get to, there is a recessed area near the pylon base with a corrugated surface that can't be sanded (fig. 4-15). Completely remove the kit corrugation and add some railroad clapboard building siding with a pattern of .020. Reduce the thickness of the pylon so the depth of the shim matches the appearance of the original, and glue a narrow strip of this onto the inside of the pylon (fig. 4-16).

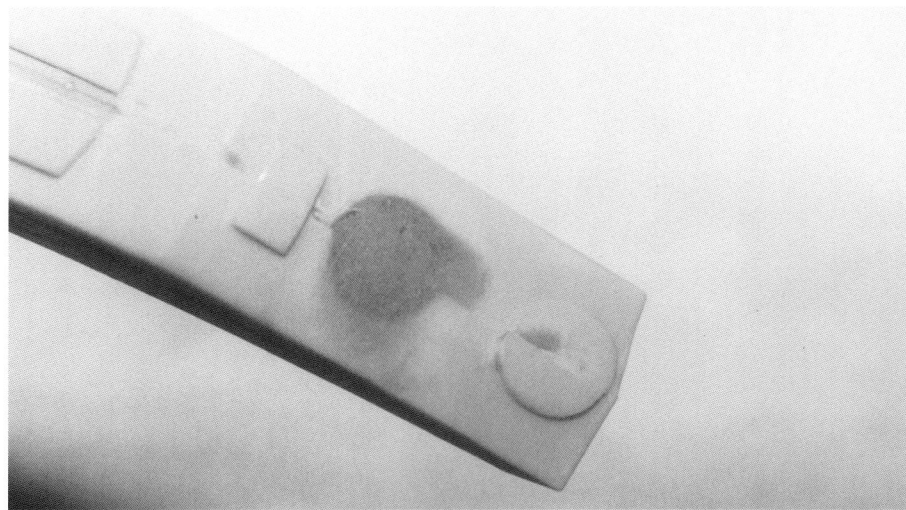

Fig. 4-10. Fill the depression in the upper pylon.

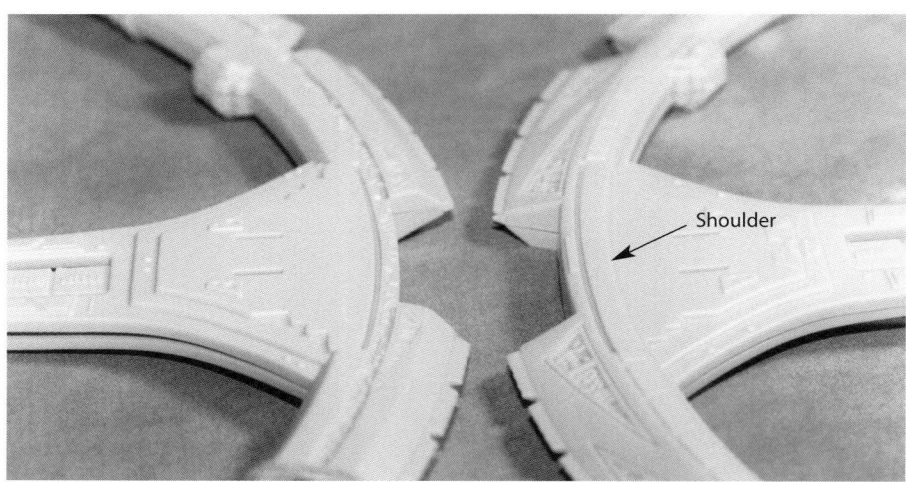

Fig. 4-11. The top of the docking ring (right) has a shoulder that is missing on the underside of the ring.

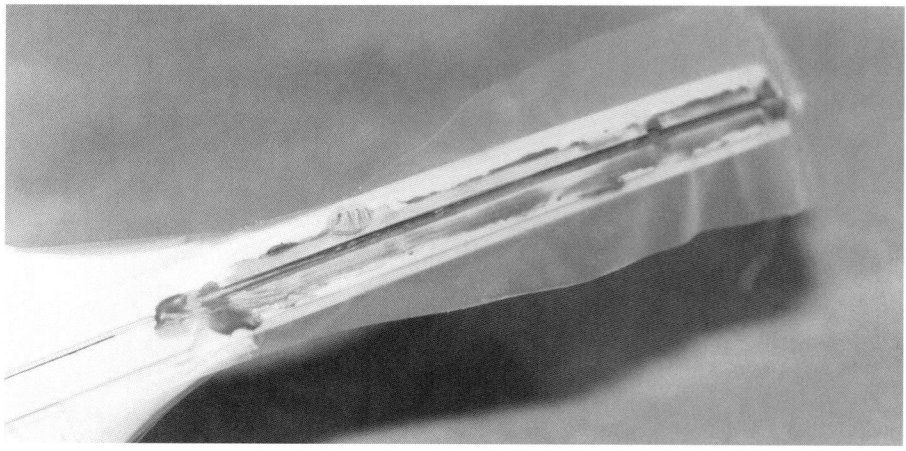

Fig. 4-12. When filling the seam on the outside of the pylon, use tape to protect the area around the seam.

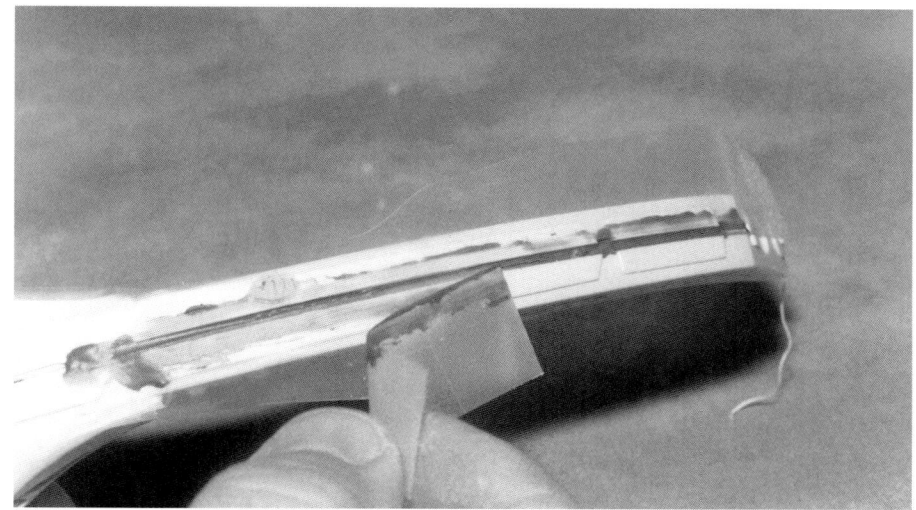

Fig. 4-13. Peel back the tape and let the filler dry.

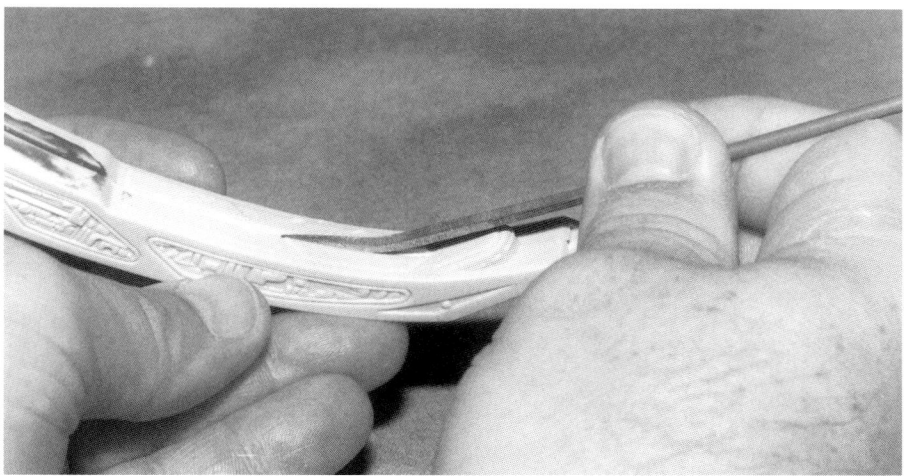

Fig. 4-14. The seams on the inside of each pylon will test your patience. To remove the filler, use a riffler file to get the most surface contact.

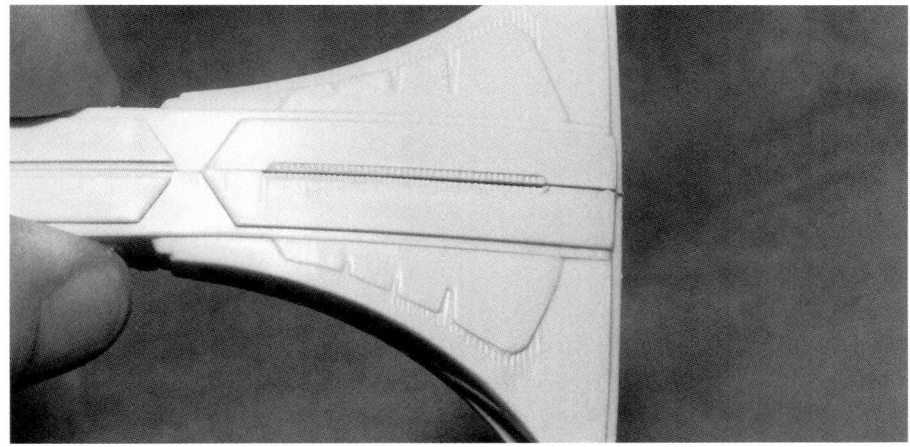

Fig. 4-15. Filling the seam and retaining the detail is impossible.

The pylons do not fit very well onto the docking ring, either. There is a substantial gap on the inside of each pylon-ring joint and a lesser one on the outside. Again, the outside seam is easy to get to while the messy one is the hardest one to repair.

Position a pylon using rubber bands or tape, and examine the gaps. Test-fit various thicknesses of strip styrene into the gaps until you find one that comes closest to matching the gap. Then glue shims of that size to the bottom of the pylon (fig. 4-17). All of the pylons on the top can be treated the same, but repeat the test for the bottom pylons. After sanding the edges of the shims to blend them into the shape of the pylons, attach them to the docking ring. Again, be generous with the glue around the mounting tab. Fill the remaining gaps at the bases of the pylons. If you find sanding or filing the putty is too difficult because of the close quarters, gently scrape the excess filler with an X-acto knife until you get a good edge.

The command center on top of the hub and the reactor on the bottom are subassemblies that fit rather well. The reactor core has a clear red piece (fig. 4-18) that provides a good opportunity for lighting to create the glowing effect you see in the show. Before putting it into place, paint ribs onto the window areas and the spike. These are visible in the show (especially the opening credits) and on the box art, but are not provided for in the kit instructions. Start by placing a piece of drafting tape that is the width of the rib down the center of each pie-shaped window.

Place pieces of tape on either side of the first piece so that the entire window is covered. Remove the center strip to expose the area to be painted (fig. 4-19). Run tape down the sides of the spikes to protect the clear areas there and trim the ends to form the area for the cap.

There are two minor problems with the command level. Part 20 has very pronounced ribs all around it (fig. 4-20). Not only do these seem oversized, their positions do not provide enough room for the distinctive oval window in Captain Sisko's office. My solution was to remove one set of ribs to create the space for the window (fig. 4-21). The overall command center seems a little undersized when it is put into place. This impression remains when you look at the real thing (can you say that about a sci-fi subject?) in episodes and compare relative sizes.

Painting

Compiling the information on the paint scheme is a problem. The instructions merely refer you to the box art, which is a stylized, metallic gray or brown, allowing a lot of room for interpretation. The side panels of the assembled model are useless—they are in some outlandish tan scheme. The best source of colors I found was the opening titles of each episode. Scenes from the show are photographed under strong, directed light, which I believe is the source of the brownish shades. There you will also find additional details to add to the model, such as the strobe lights for the pylons and contrasting colors in some of the panels.

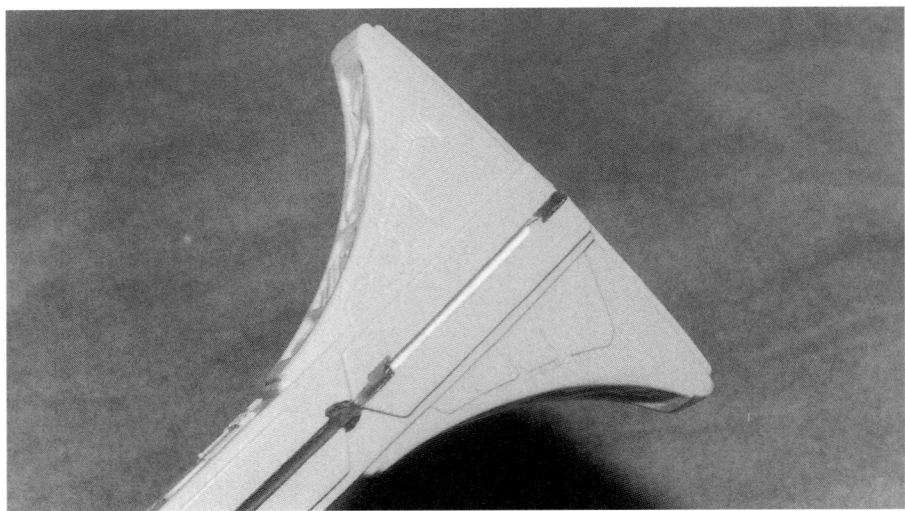

Fig. 4-16. Replace the entire area with corrugated sheet plastic.

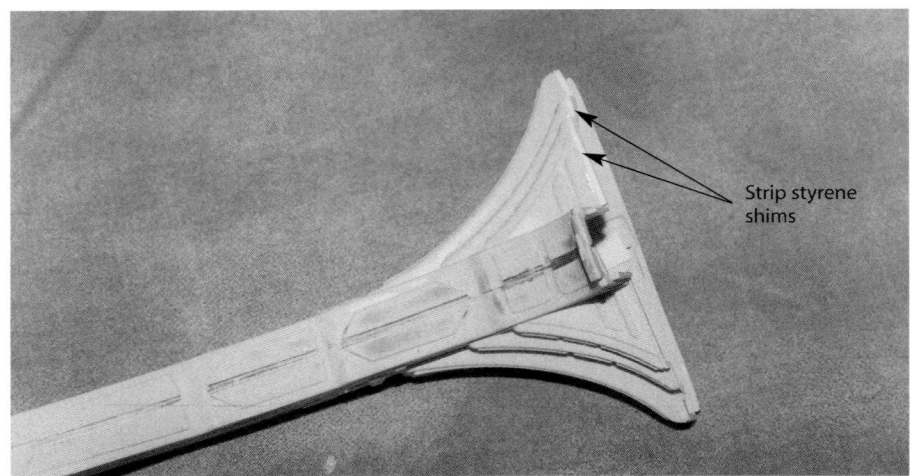

Fig. 4-17. Add shims to get a good fit on the inside of each pylon.

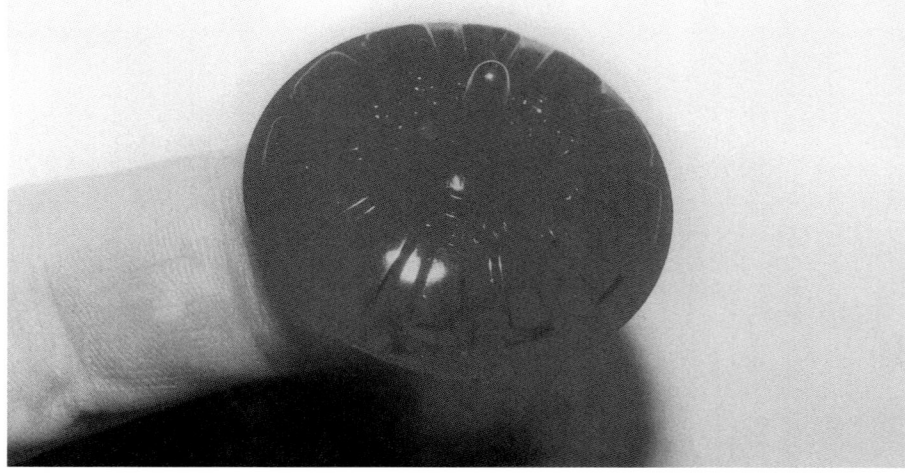

Fig. 4-18. The reactor core is a clear piece that needs some additional detailing.

Fig. 4-19. Mask the reactor with drafting tape to spray on the ribs.

Fig. 4-20. The command level has prominent ribs molded on the part.

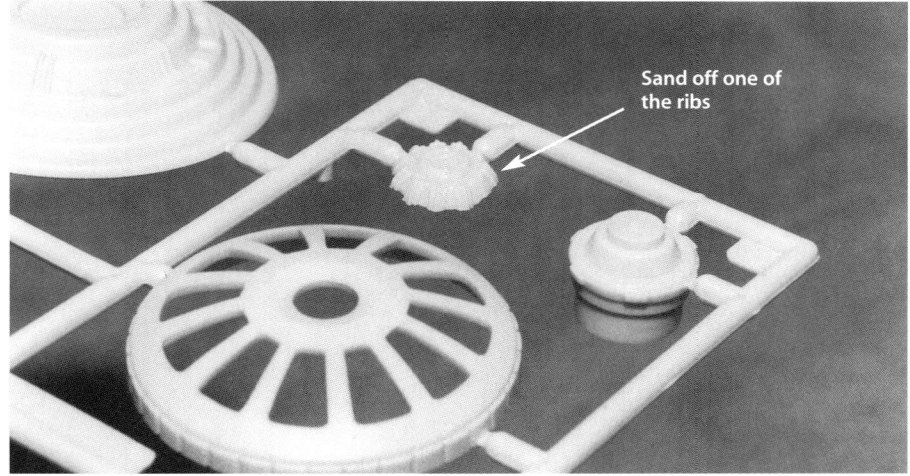

Fig. 4-21. Remove one set of ribs from the command center to make room for the commander's office window.

I chose to ignore any brown paint and used Testors Metalizer paint to get the metallic effect apparent on the show, with an overspray of Dullcote to eliminate much of the sheen. No one color out of the bottle appeared to be the right shade. I chose a 50/50 combination of Titanium and Magnesium. A wash is also a must to bring out all of the surface detail. If you want to superdetail the paint, you can add many panel lines, using scenes from the show as your guide. To get a third layer, drybrush the high points of the details with a lighter gray, such as Dark Ghost Gray. This also enhances the "weather-beaten" appearance, in keeping with the overall condition of the station.

A couple of painting details aren't covered in the instructions but help dress up the model. Paint the grilles on the sides of the sails a dark red (fig. 4-22). The triangular panels on top of the docking ring should be a mustard yellow. The mustard should not be a solid color, but should be less intense on the raised areas of the panels. To control this, I used watercolor paint from a standard children's paint set. I layered in the color, letting it accumulate in the recessed areas while it remained very thin on the peaks. Each coat added more color until I achieved the intensity I wanted (fig. 4-23). Where I overdid it, I gently rubbed the high points of the panel to remove the excess paint.

The main docking port lights are located in flat, circular areas at the base of each pylon. Lightly drill out the area, paint it silver, and add a drop of clear orange on top (fig. 4-24). You can make an even more dynamic light using small, smooth-lensed railroad lights with the clear orange painted on the lens.

For a little added detail, paint one of the runabouts included in the kit and place it in the docking bay (fig. 4-25).

The most time-consuming step is adding all the windows. On the central core, many are represented by very tiny dots (much like the Romulan Warbird kit in the Adversary set). Most are totally overlooked. Again, use the opening credits to develop a feel for the way you want them to look. Rather than try to be precise, I felt that an alternating pattern of four windows stacked in two rows would carry the feeling without being overly complicated to apply.

To add the windows, use a paint pen, which you can find in the art supply section of a craft store. The one I used is made by Sakura, but there are other brands as well. Based upon the alcohol odor, I assume they contain an acrylic paint. As before, the use of different mediums is important. If you make a mistake, you can clean off the acrylic paint with alcohol and not disturb the hull color underneath.

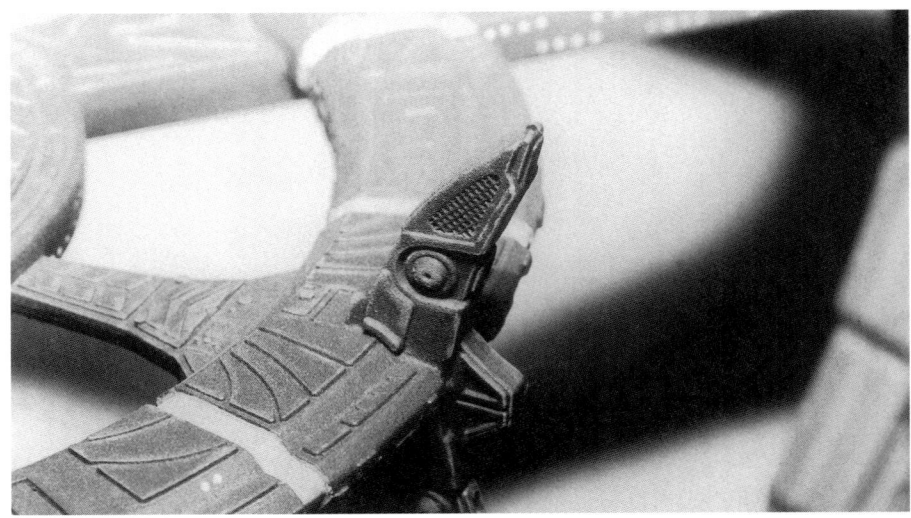

Fig. 4-22. Paint the detail grille red and add the strobe to the top and bottom of each sail.

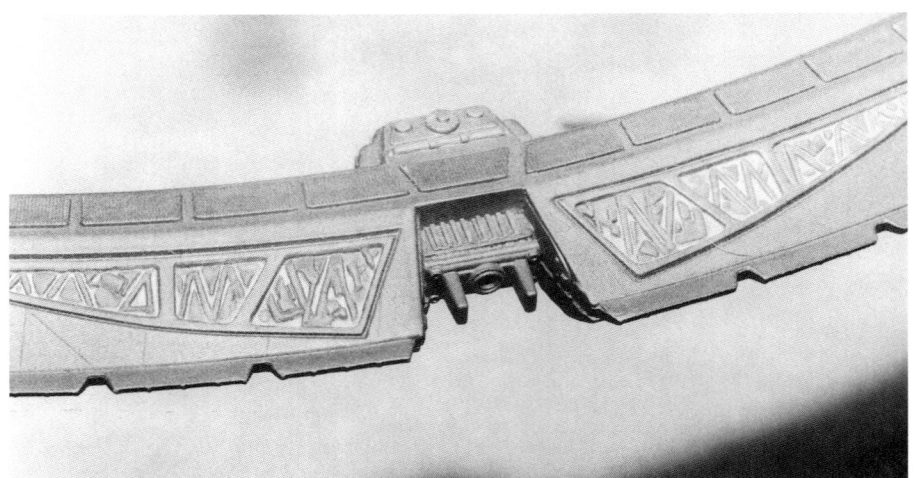

Fig. 4-23. The background color for the accent panels can be applied as a wash with a water-based paint.

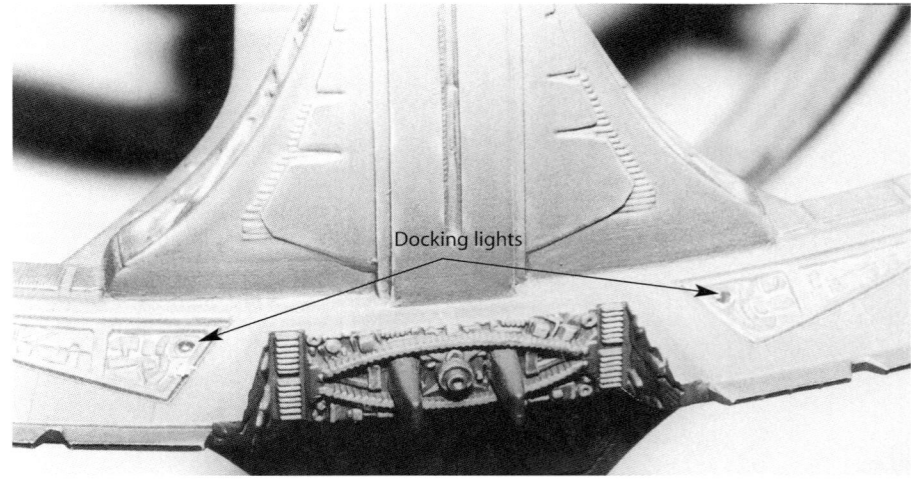

Fig. 4-24. You'll have to add docking lights to each side of the main docking ports.

Fig. 4-25. There's even room to add detail to the 1/2500 scale runabout.

Fig. 4-26. Apply the windows using a white paint pen.

White is slightly more difficult to find than other colors. The smallest point I found was a 1mm. It was adequate, albeit a little large, for this model. You could get smaller dots by carving down the point. Brace your hand against a prop or even the hand holding the model and slowly add the pattern along the inside of the spokes, the outside of the habitat ring, and the inside of the docking ring (fig. 4-26).

The completed model is imposing, especially considering the small scale, with an overall diameter around 16″.

5
Lighting an Ambassador-Class Starship

From the time *Star Trek: The Next Generation* started, many fans speculated about the appearance of the two "missing" ships in the series. What did the *Enterprise-B* and *Enterprise-C* look like? It wasn't too difficult to imagine that the -B would be a variation of the *Excelsior* class, but the -C fired the imagination. For a while, several people produced artwork speculating on the appearance of the ship, but the "official" word didn't come until the Next Gen episode "Yesterday's *Enterprise*" aired in February 1990. Interestingly, even though "Yesterday's *Enterprise*" was considered by many to be the best episode of the series, *Ambassador*-class starships have been rarely seen or mentioned. Perhaps this explains why it took so long for the -C to be released in kit form? AMT/Ertl released the kit of the *Enterprise-C* in late 1998.

In 1999 a second kit was released. This one contained a light kit and the markings for two additional ships—the USS *Yamaguchi* and the USS *Excalibur*.

Fig 5-1. A unique way to light a kit!

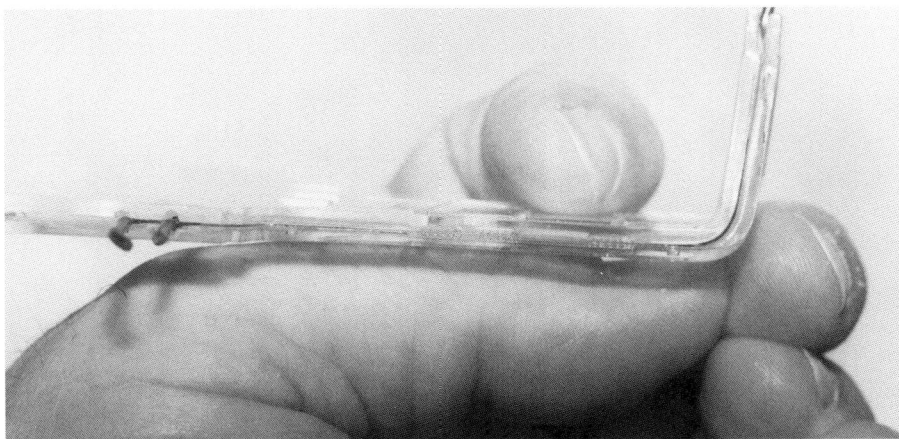

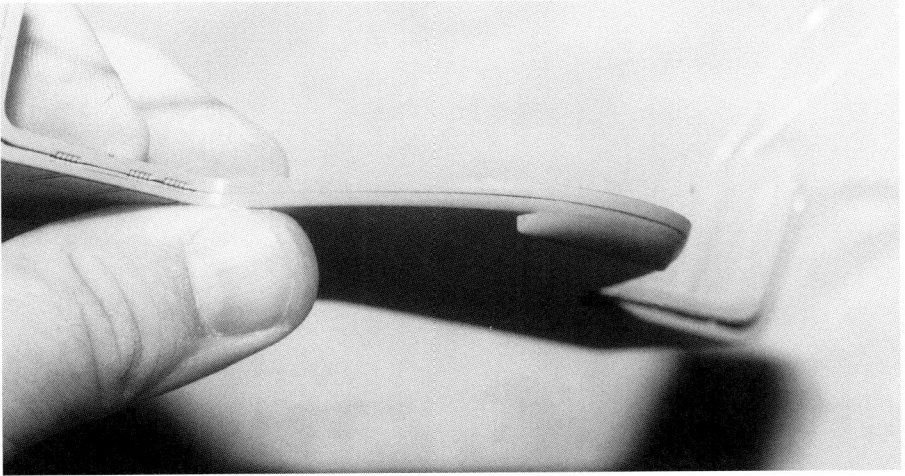

Figs 5-2 and 5-3. Paint or prime the parts early in the process to make it easier to find the seams that need to be filled.

The *Yamaguchi* was referred to in the DS9 episode "Emissary" and was the ship Benjamin and Jake Sisko traveled aboard when they first came to the station. *Excalibur* was commanded by Will Riker during the Federation attempt to prevent Romulan aid to the Duras family during the Klingon civil war in the episode "Redemption Part II."

If you have seen the earlier lighted kits, you are in for a shock when you open the box. Instead of the standard molded styrene parts, this kit is made up entirely of clear parts (fig. 5-1)! This approach to making a lighted kit, along with the electronics themselves, forces you to take a very different approach to building the model.

Painting issues

The first question is when to paint the model. The instructions recommend waiting until the model has been fully assembled before applying an overall coat of medium gray, then applying the detail colors. I would strongly suggest that you apply the base coat as you complete each subassembly. This acts as a primer coat and highlights the seams for filling (figs. 5-2 and 5-3). (This is a *Star Trek* kit—you know there will be seams!) In the same vein, use the typical approach for most models and paint the details on the subassemblies as you go because they are easier to get to while they're still on the sprue.

This approach also helps protect the ultimately clear parts from overspray. As you put together the subassemblies, you will have to fill and respray seams, but this limited painting will be easier to control than the overall application of paint as a final step.

The biggest problem with the *Yamaguchi* kit is the painting instructions. As I mentioned, the instructions call for an overall base color of medium gray, which generally means something around the 36300-36200 range on the FS scale. The patterns are a mixture of 90 percent or 80 percent Insignia White (FS33637) and 10 percent or 20 percent Insignia Blue (FS35044). The result is a very dark model with the "dark" accent panels lighter than the medium gray base color. (Refer to Appendix 1 for some thoughts on mixing paints.)

Most modelers will use either Insignia White or Camouflage Gray (FS36622) as a base color to get a scale effect for the model. This time I chose white. The recommended mixtures for the accent panels can then be used to achieve the proper appearance.

A word of warning. Because the model is molded in clear plastic, you will need additional layers of paint to make the model sufficiently opaque so the lights won't show through the hull. As you paint the engines and secondary and primary hulls, use a penlight to test the finish (fig. 5-4). If you can see the light shining through, the kit lights will also show through. The simplest way to solve this problem is to paint the inside of the model with a heavy coat of gray paint. Be sure to leave holes for the light to shine through wherever there will be lighted windows.

Fig 5-4. You must keep any light from shining through the hull.

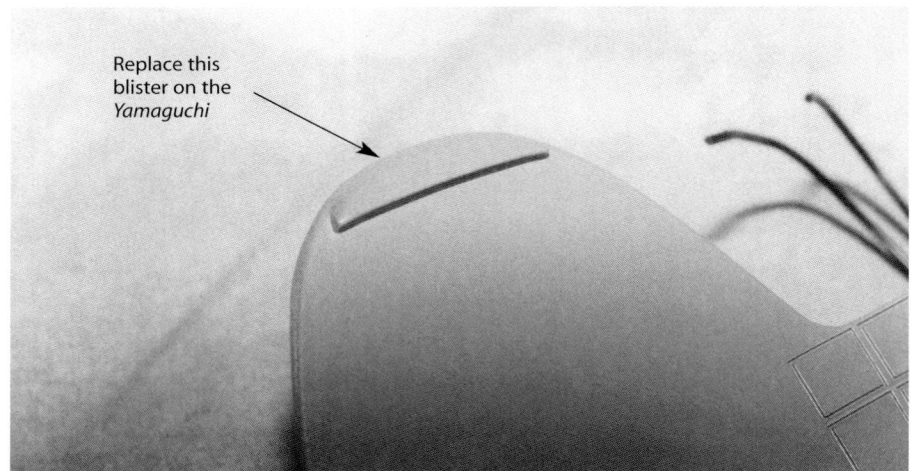

Fig 5-5. Remove or not? Decide which ship you want to build early in the construction process.

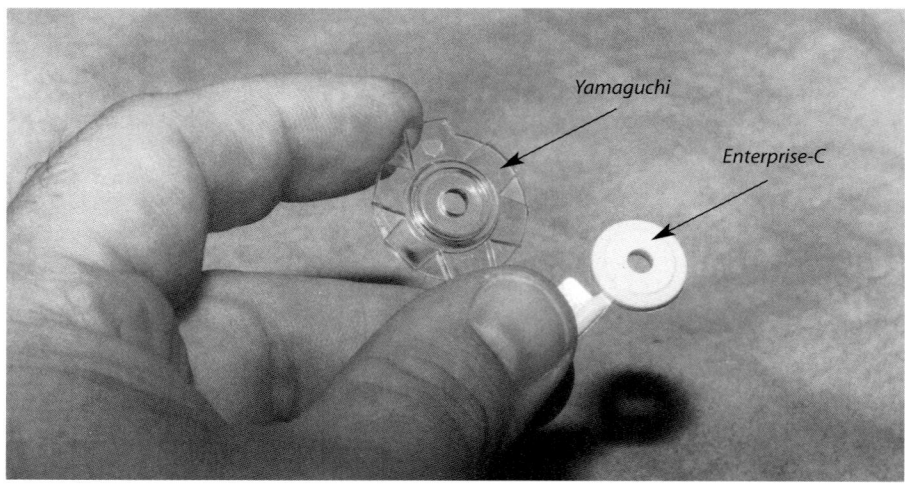

Fig 5-6. Replace the part from the *Yamaguchi* kit with the one from the *Enterprise-C*.

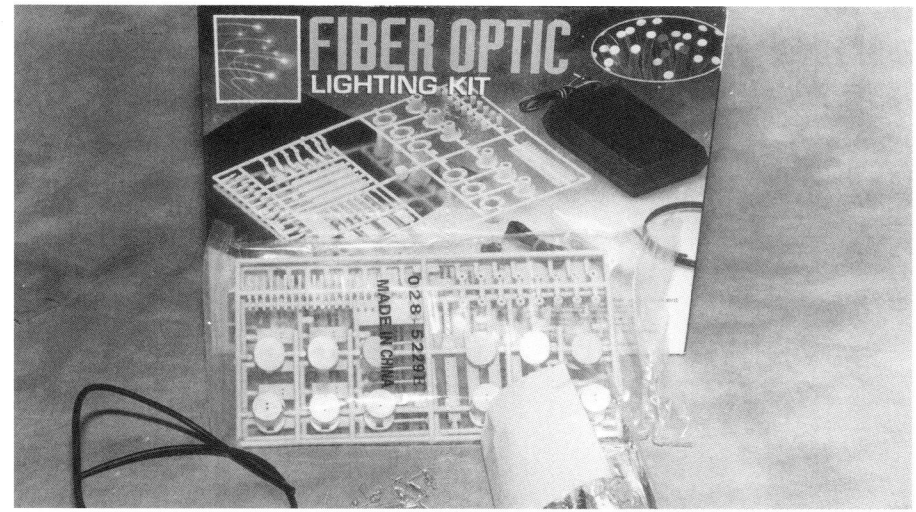

Fig 5-7. A generic light kit is also available for lighting other models.

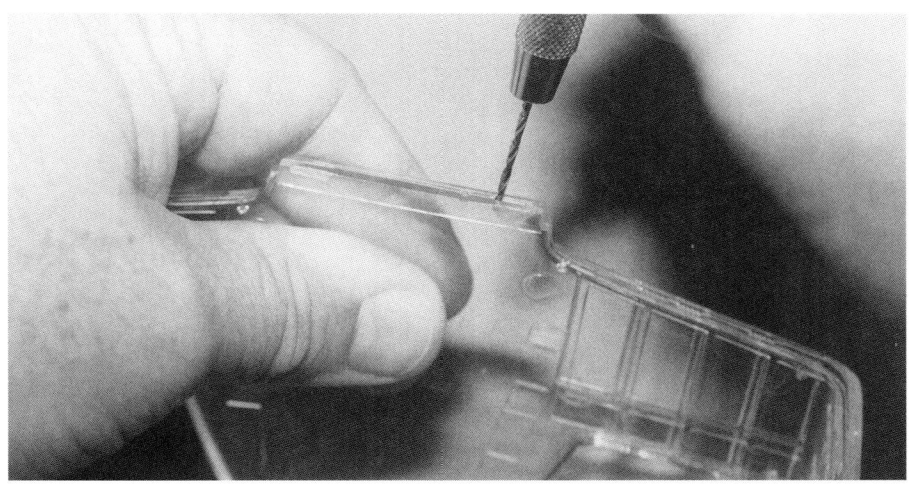

Fig 5-8. Align the grooves in the parts by drilling a pilot hole.

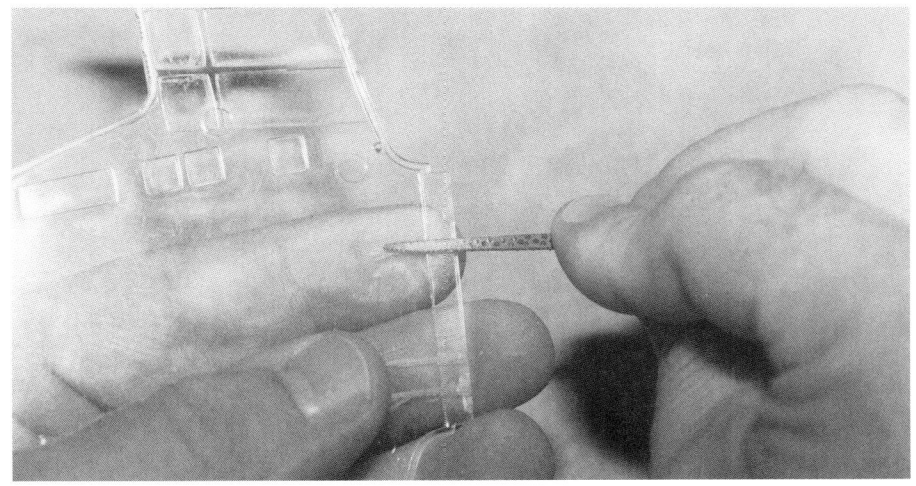

Fig 5-9. Cut grooves to run the wires through the bordering lip.

Which kit to build?

To build a *Yamaguchi* or *Excalibur*, you must make some physical changes to the model. Changes were made in the studio model from the configuration seen in "Yesterday's *Enterprise*" to the configuration seen in other ships in later episodes. The *Yamaguchi* molds are (almost) the same as the *Enterprise-C*, but you must do some surgery to change the model into the later variant (fig. 5-5). This consists of removing some of the escape pod blisters, adding new ones in different locations, and adding an enlarged blister below the shuttle bay.

If, on the other hand, you have both kits and want to make a lighted *Enterprise-C*, use those decals and ignore the *Yamaguchi* changes. The only external, physical change you will make is to substitute part 3 from the *Enterprise-C* kit for the underside of the main hull (fig. 5-6). Refer to the painting instructions from the *Enterprise-C* as well, since the pattern is slightly different.

Don't forget to keep the instructions and extra parts from the *Yamaguchi* kit so they can be used to build a nonlighted version.

Step 1 – The light kit

The power box, lights, and wires are customized for this kit. A generic light kit, also produced by AMT, is available separately (fig. 5-7). This version contains additional parts so other kits can be lit, perhaps even more than one. Only one battery box and switch are included. They aren't particularly special, so you can buy a substitute at an electronics store.

Step 2 – Main engine pylons

Notice that the upper and lower pylons (parts 8 and 9) have raised lips, which are the contact points when they are glued together. According to the instructions, you are to cut four grooves in the bottom plate (part 9). These will allow the wiring running from the engines to the secondary hull to pass through.

Even though you are instructed to strip the insulation from the wires that run through this area, space becomes a problem. I also added a groove to the upper pylon plate (part 8). To make sure the grooves line up, hold the parts together and drill a pilot hole for each wire centered on the joint (fig. 5-8), then use the scored areas in each part to guide a file for making the rest of the grooves. Even though it isn't mentioned in the instructions, you should do the same thing at the other end where the pylons join with the engines, because the lips also run through this area. Use a needle file to make the groove large enough for the wire (fig. 5-9).

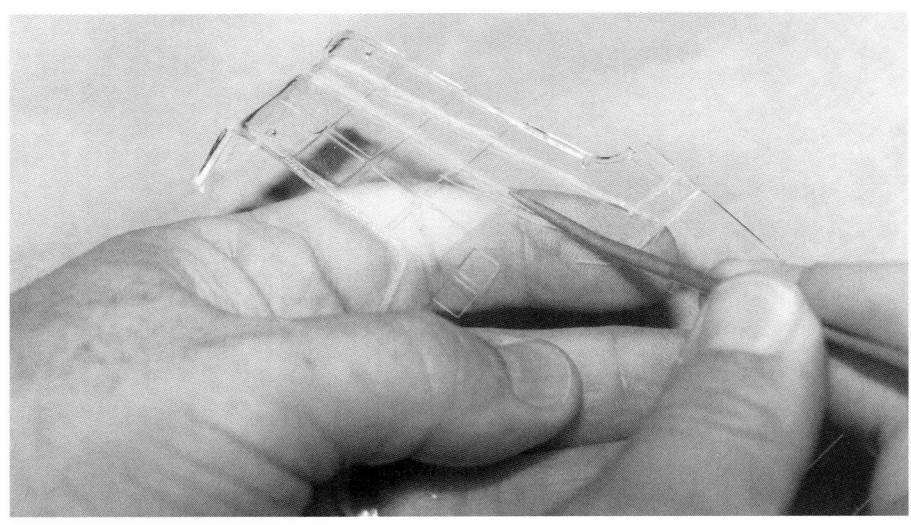

Fig 5-10. Channels sanded into the parts add extra space for the wiring and keep it from shifting.

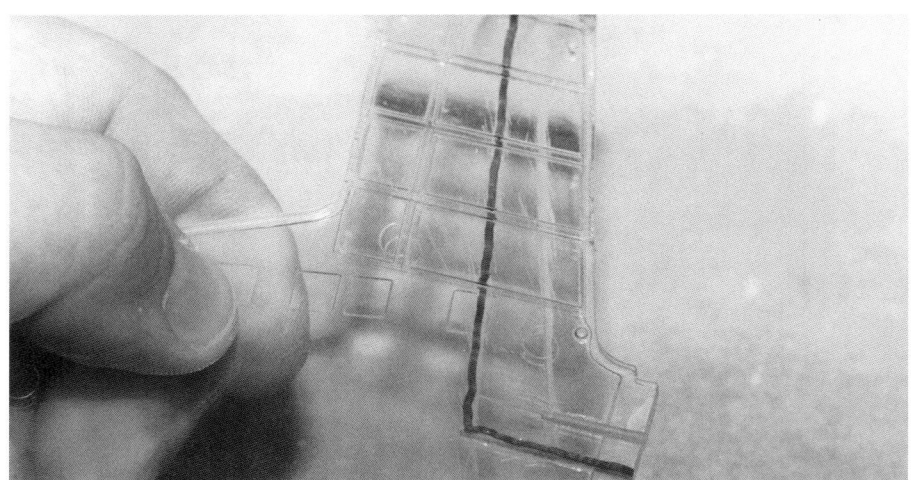

Fig 5-11. Trace a line on the second part following the channels cut in the first.

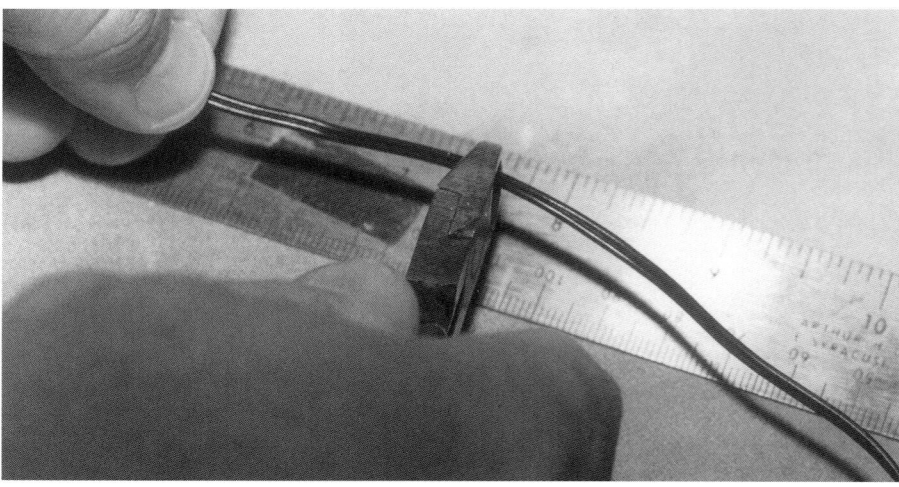

Fig 5-12. Measure the wire according to the instructions and cut it to length.

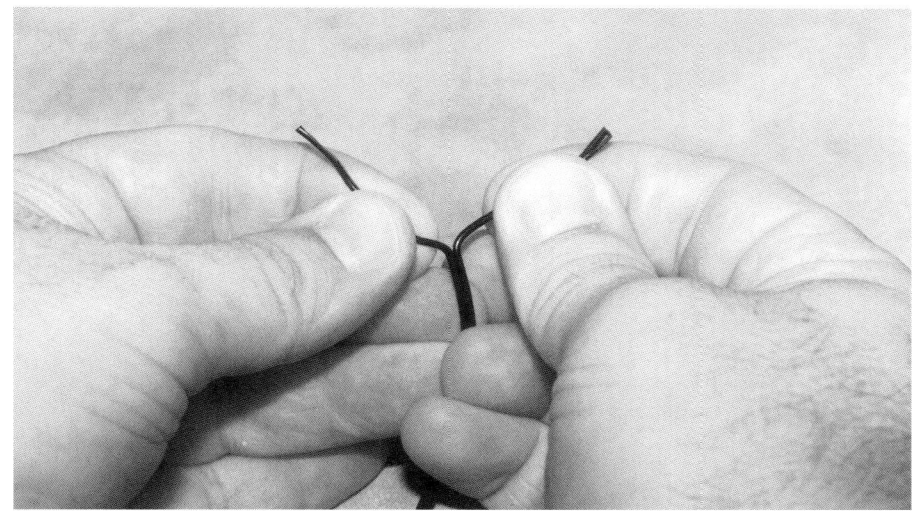

Fig 5-13. Split the wire into individual strands.

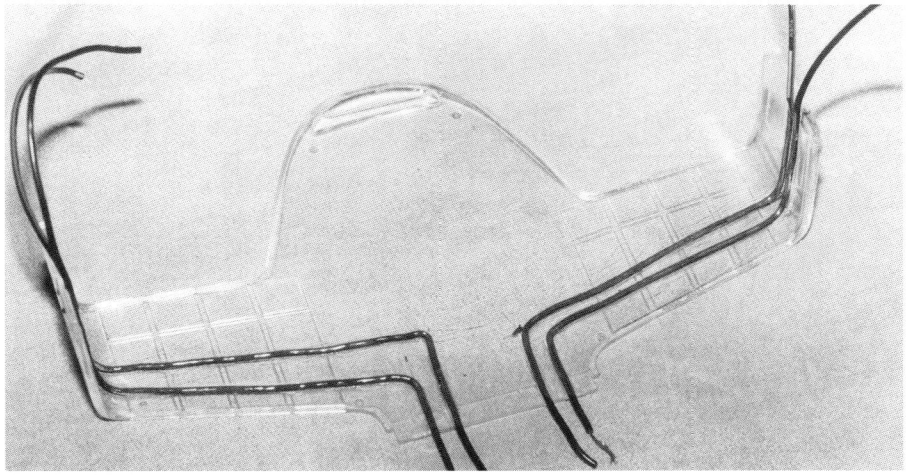

Fig 5-14. Glue the wires into one part before sandwiching them in.

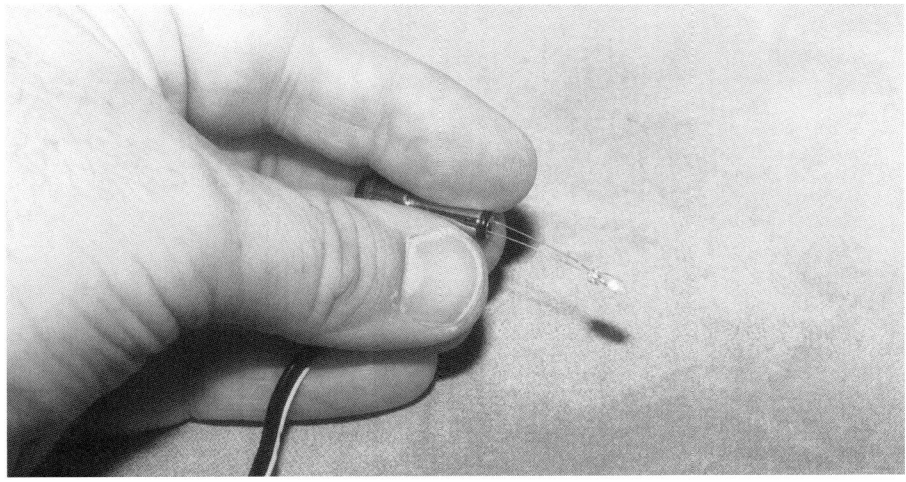

Fig 5-15. Test the bulbs before using them.

The area between parts 8 and 9 has a hollow space for the wiring to run through. The instructions say to strip off the insulation so it will fit in the space provided. I had no luck in doing this, snapping several copper strands as I tried to remove the insulation. Plan B was to put in the wire, insulation and all, but this required me to cut channels in both parts (fig. 5-10) to create enough extra space for the insulation. To make the channels line up, cut a channel into one part, then trace its path onto the outside of the other part (fig. 5-11) and use the tracing to add the channel to the inside of that part. The channels also make it less likely that the wires will come loose and touch, causing a short circuit.

Follow the instructions and cut the wire to the prescribed lengths (fig. 5-12). Split the wires into individual strands (fig. 5-13) and super glue them into the channels (fig. 5-14).

Step 3 – The secondary hull

The secondary hull is the first place to add the bulbs. Test each bulb as shown in the instructions before attaching it to the wiring (fig. 5-15). I was a little nervous about using a soldering gun around styrene (although it is possible), so I tried soldering strips from Radio Shack. These are low-heat strips (fig. 5-16) that you wrap around the joint. You then pass a lighted match above a strip until it flows over the joint. Twist the ends of two sections of wire with one lead from a bulb and solder the connection (fig. 5-17). Twist the bulb so the leads won't cross and let the stiffness of the wire maintain the separation. Using cellophane tape to insulate the exposed wiring is another low-tech trick you can use on this kit (fig. 5-18).

Before starting the hull assembly, do the preliminary painting and filling. There are sink marks along the aft underside of the hull and above the location of the impulse engine that need work. Once you have smoothed out these areas, apply the base color to the hull. Paint the detail to each half of the hull as well.

After assembling the string of lights, place the two sides of the hull together and slide the pylons into place, but glue only one side of the hull to the pylons. Remove the loose side (fig. 5-19) and finish the internal assemblies. Run the main power line through the sleeve for the base and the other end up the neck to the saucer section, install the impulse engine, and attach the main deflector dish.

Fig 5-16. Solder strips are a low-tech way to join the wires.

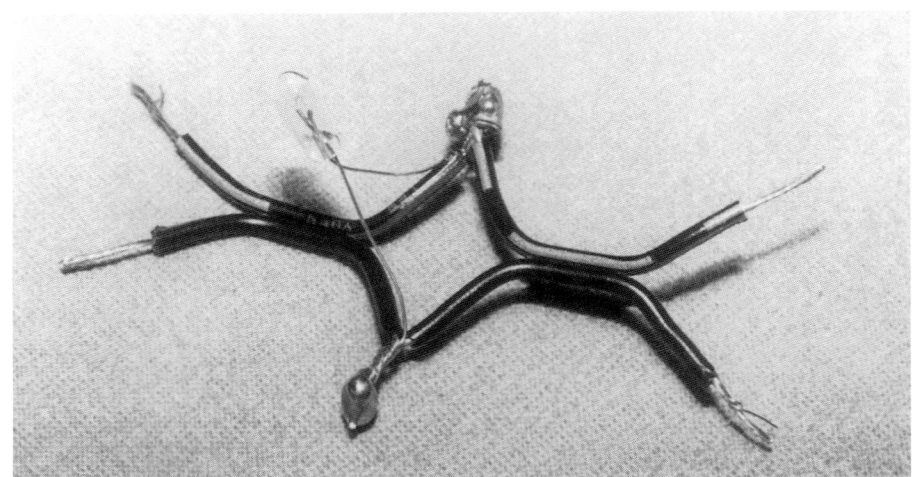

Fig 5-17. Make sure the "legs" of the bulb don't touch and cause a short circuit.

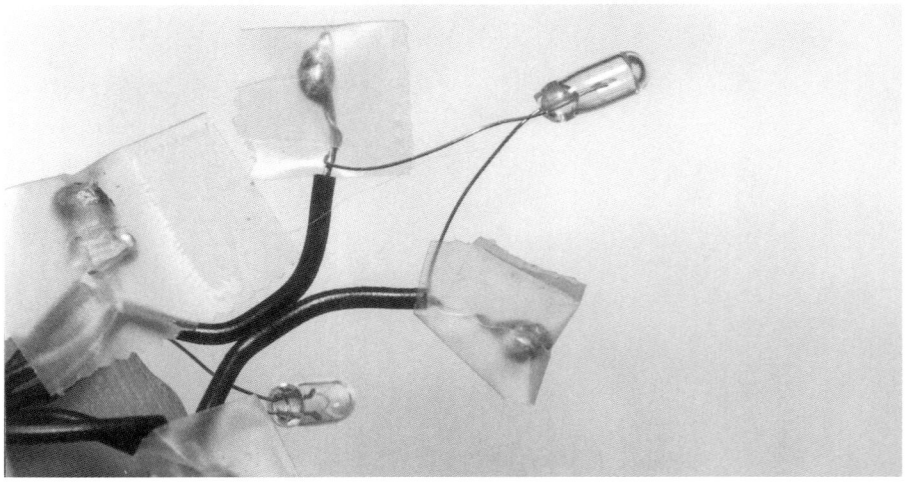

Fig 5-18. Since this is a low-heat, low-power arrangement, tape is adequate to insulate the exposed wire.

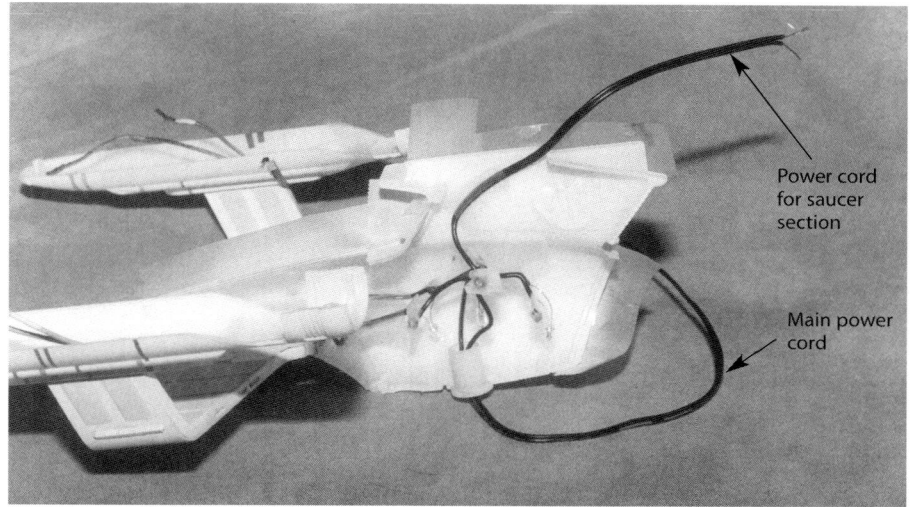

Fig 5-19. Lighting the secondary hull results in a real "rat's nest."

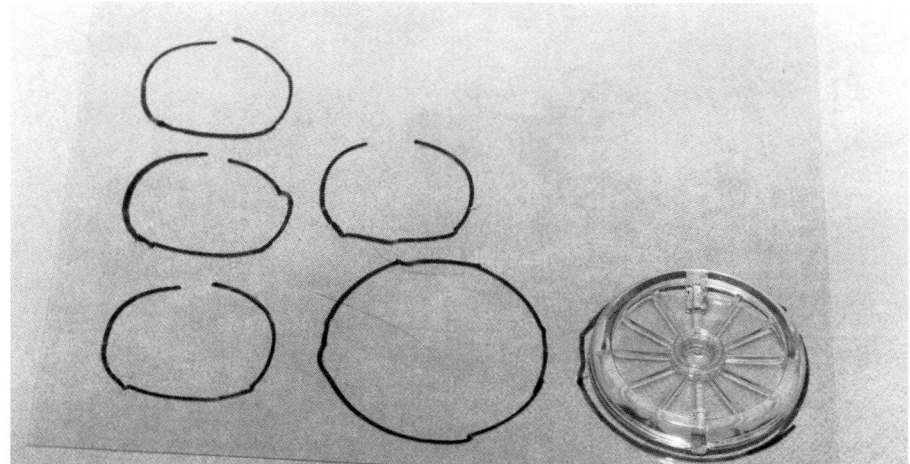

Fig 5-20. Trace patterns for the deflector and the bussard collectors onto clear sheet acetate.

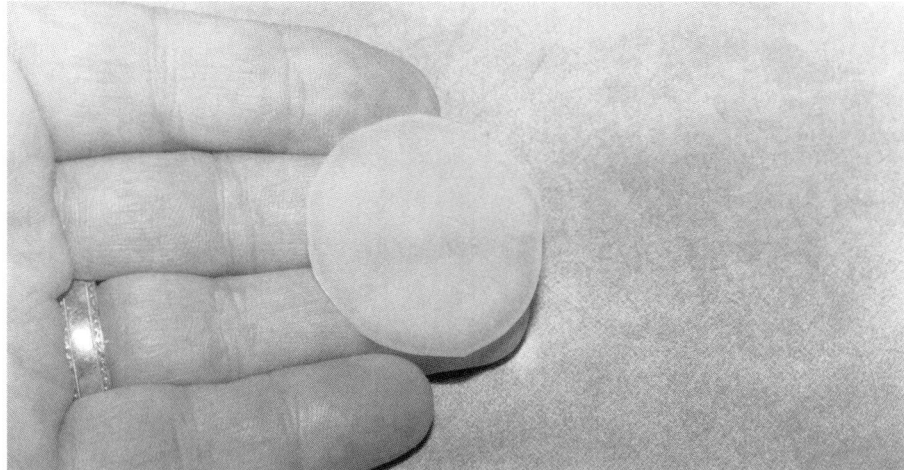

Fig 5-21. Sand the disk until it has a milky appearance.

Effectively reproducing the soft, blue glow of a deflector has always been my biggest challenge. With this model, the lights solve half the problem, but the problem how to get the gradual change from blue at the edge to nearly white in the center remains.

My approach was to take sheet acetate and, using the parts, trace disks onto the sheet (fig. 5-20). After cutting them out, gently sand them with 400-grit sandpaper until you achieve a uniform translucence (fig. 5-21). Airbrush clear blue onto the disk (fig. 5-22), and then sand the center of the disk until the paint has been removed (fig. 5-23). Make more than one so you can practice and have a chance to mess up.

Glue the disk into place behind the location for the dish (fig. 5-24). For the dish, use an adhesive like Kristal Klear to prevent crazing the finish. Eventually, the dish will receive a coat of flat, along with the rest of the model, further softening the glow.

Once the glue has dried, gently position the bulbs to get the maximum lighting effect. Position one of them so it is centered behind the disk. Glue the opposite side of the secondary hull into place. After painting the rim of the deflector, glue it into place (fig. 5-25).

Now comes the fun part of filling and sanding the joint for the hull halves and the joint between the hull and pylons. Careful masking of the already painted areas is a must. The worst area to fill and sand is around the plate for the impulse engine (fig. 5-26). As usual, the most difficult place to work is the spot needing the most work.

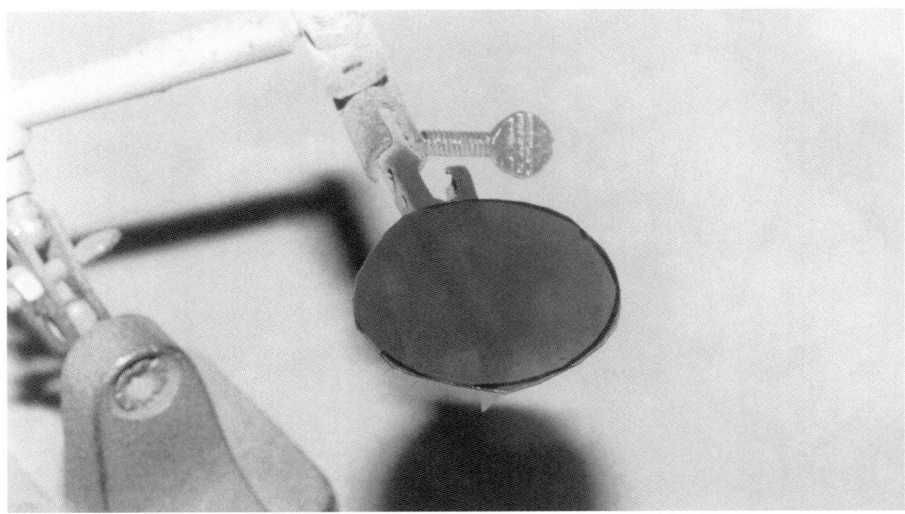

Fig 5-22. Apply translucent blue to the disk.

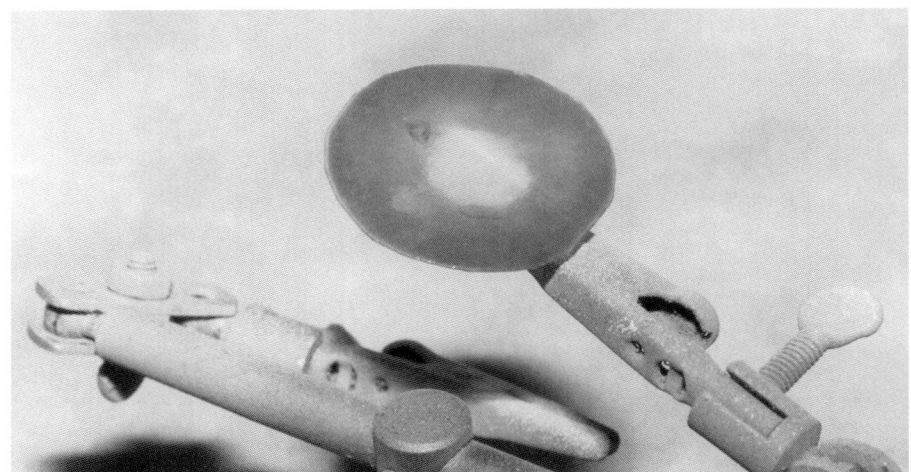

Fig 5-23. Sand away the paint from the center of the disk.

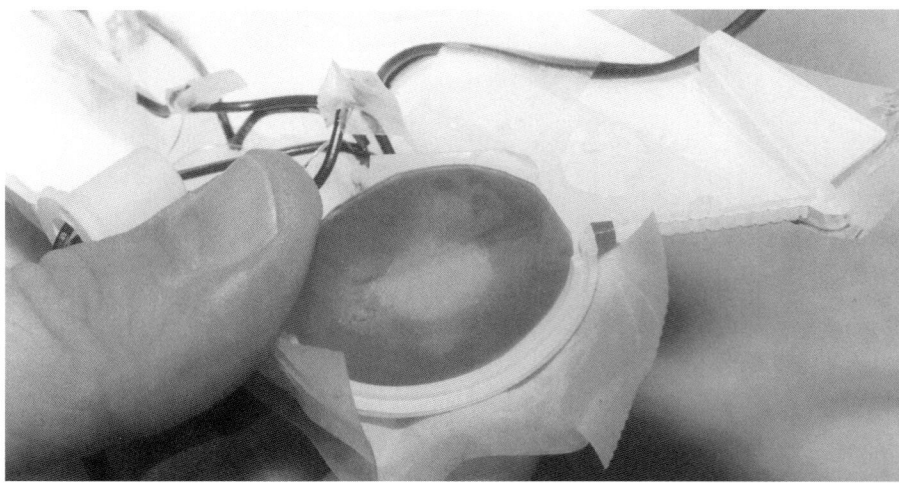

Fig 5-24. Glue the disk behind the ring where the deflector dish will be located.

Fig 5-25. Mount the deflector and fill the seam around it.

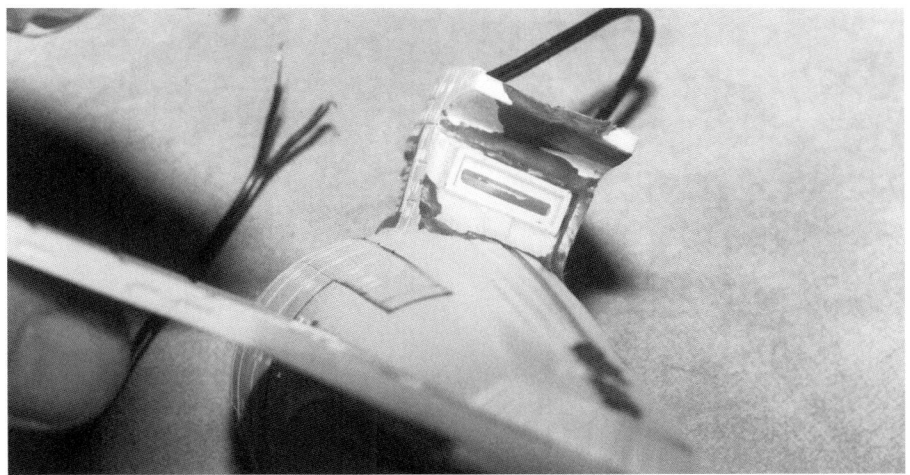

Fig 5-26. The hardest seam to get to is always the worst one on the model.

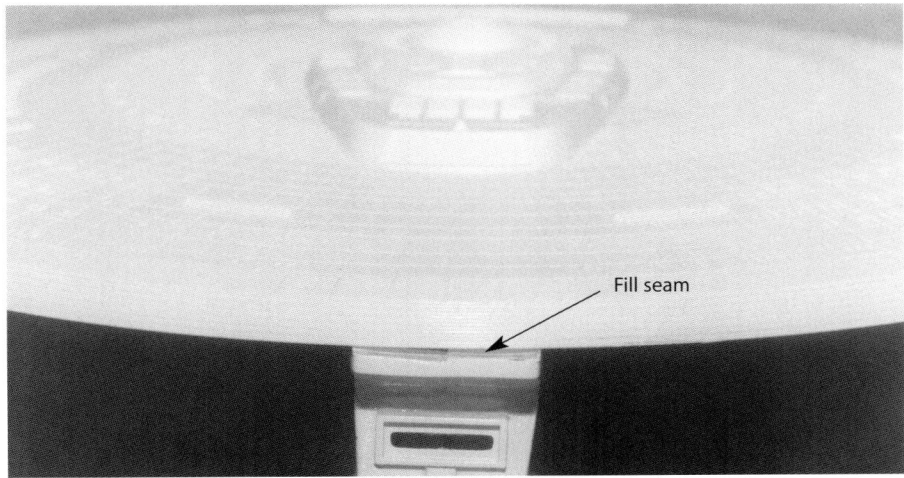

Fig 5-27. Before attaching the top of the saucer, fill the gap on the "boattail."

After that is finished, check the "boattail" extending behind the saucer. Fill this area and sand it flat to create a clean joint (fig. 5-27). Finally, pay special attention to the corrugated seam on the neck. There is a substantial mismatch, and the only way to camouflage it is to fill the entire area and cut new grooves (fig. 5-28).

Step 4 – The warp engines

Cut holes in the bottom of the engine nacelles to allow the wires to enter the engines (fig. 5-29). Even though the engines themselves are symmetrical, they *are* "handed." The port nacelle has a notch at the front of the locating slot, while the starboard nacelle has a notch in the middle. Some light sanding on the pylon tab may be necessary to get it to slide into the slot.

The engines are completely hollow, so the lights that you place inside provide lighting for the bussard collectors as well as for the engines themselves (fig. 5-30). This works fine for the engine, but the collector's light should be diffused somewhat. To achieve that look, add a frosted filter like the one you used behind the main deflector. Locate it just inside the opaque part of the engine nacelle behind the bussard collector (fig. 5-31).

The clear parts of the warp engine have a seam at the rear that you'll have to eliminate. Set the parts into the top of the nacelle to get them aligned properly (fig. 5-32). Very carefully glue the joint by using a chemical glue like Tenax 7R rather than a super glue, which would craze the plastic and make the job tougher. White glues dry clear, but they don't have the bond strength to put up with the next step. Before starting to sand, tape the forward ends to a support to keep them from flexing and breaking the joint (fig. 5-33).

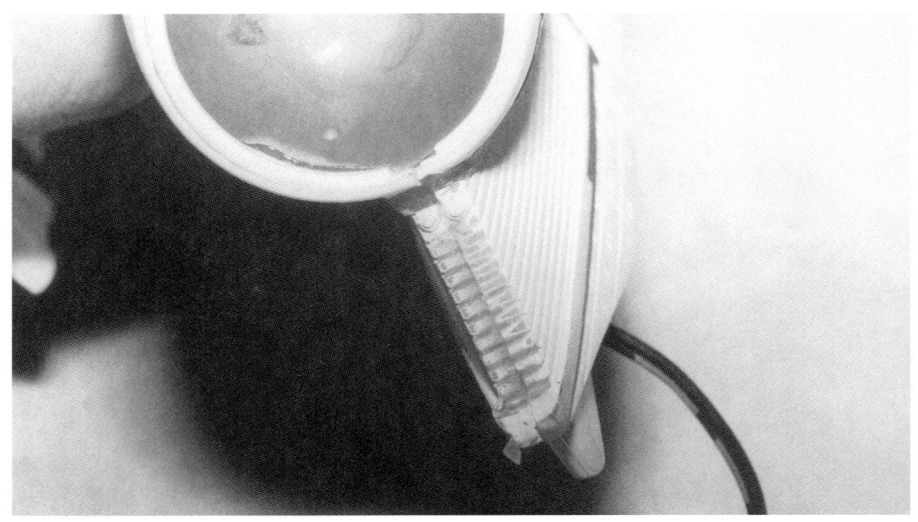

Fig 5-28. Cut new grooves on the neck to correct the mismatch.

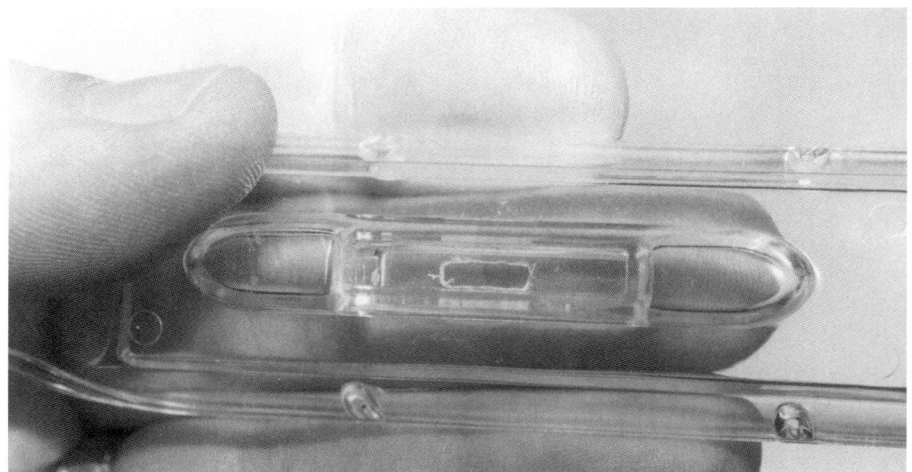

Fig 5-29. Cut a hole for the wires leading from the pylons.

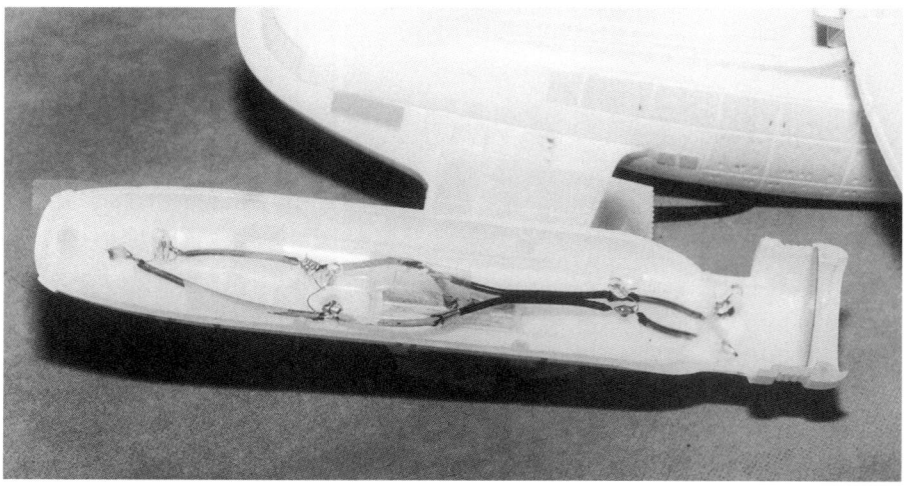

Fig 5-30. Wiring for the engine nacelles

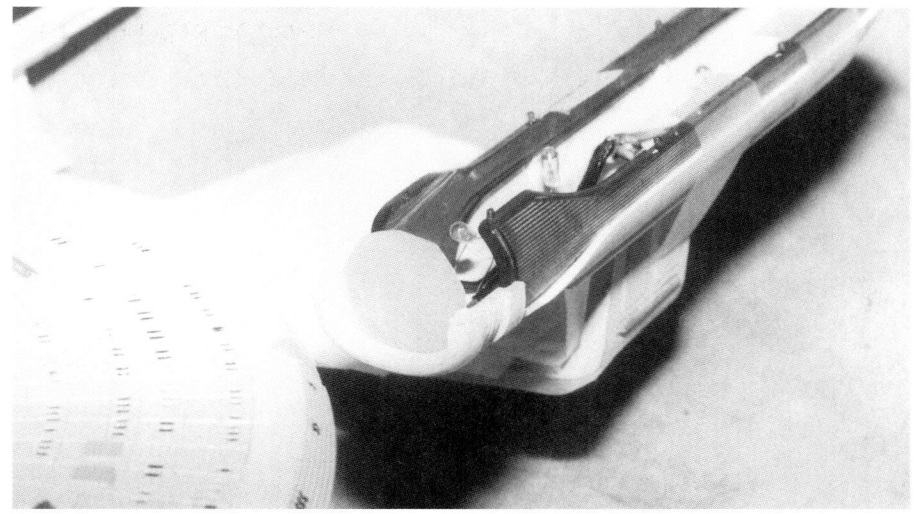

Fig 5-31. Add a diffuser to the engine nacelles also.

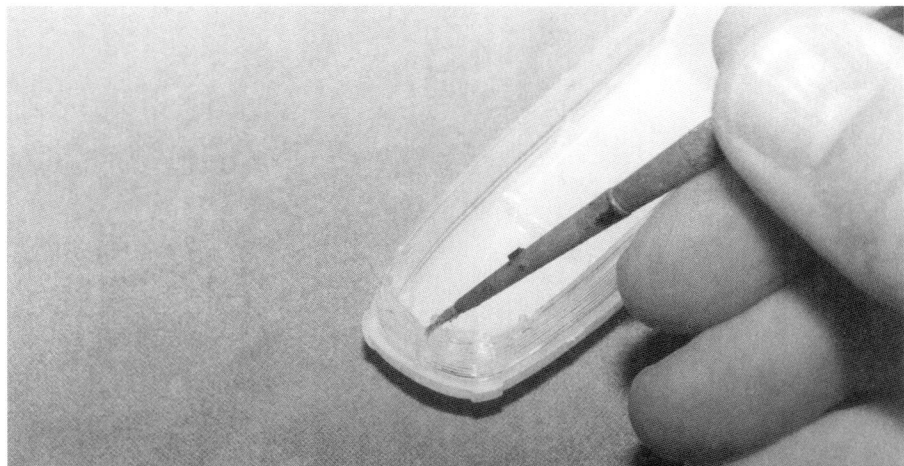

Fig 5-32. Set the clear warp engine parts in the nacelle top to align them for gluing.

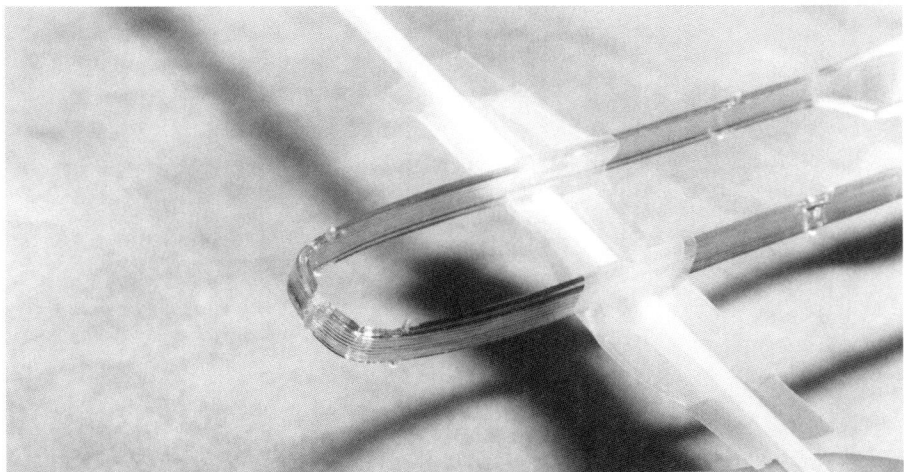

Fig 5-33. Tape the parts to a brace to take the stress off the joint while you work with it.

Using high-grit sandpaper (at least 600), carefully sand the exterior of the part to remove the seam. You can't use filler here, so you may have more sanding to do than normal (fig. 5-34).

After eliminating the seam, take a pass at the area with 1200-grit sandpaper. At this point you are eliminating the visible scratches by using progressively finer polishes. The last polish to apply is the Blue Magic compound (fig. 5-35). Some modelers use toothpaste for this step. Stop when you have a uniformly cloudy appearance.

Wash the joint to remove any residue left from the Blue Magic. Fill a small container with an acrylic floor polish such as Future. Dip the joint in the polish (fig. 5-36) and let the assembly hang, joint down, to dry. The floor polish will coat the sanded area and make it clear again.

Paint the inside of the bussard collector red and the inside of the warp Engine Blue. Now is not the time to be stingy with the paint. The bulbs provide a surprising amount of light, and too little paint will look almost nonexistent. Assemble the warp engine clear and top parts, then carefully sand and touch up the paint to deal with the seams (fig. 5-37).

Step 5 – The primary hull

After applying the base color, mask the areas that are to stay that color and apply the middle-range panel color. Use a combination of liquid mask, drafting tape, and cellophane tape—whatever best fits the area being masked. Most demarcation lines fall on the recessed panel lines in the kit, so you can use them to your advantage.

Use crepe-style drafting tape to lay down a hard edge for the mask. Crepe works best because it is easy to get it to lie flat while you're forming a curved line. Take a curved hobby blade and rock it back and forth in the recessed panel line to cut the tape (fig. 5-38). It is easier to control the pressure this way, cutting the tape without cutting into the paint or the model.

Fig 5-34. Gently sand the seam until it is smooth.

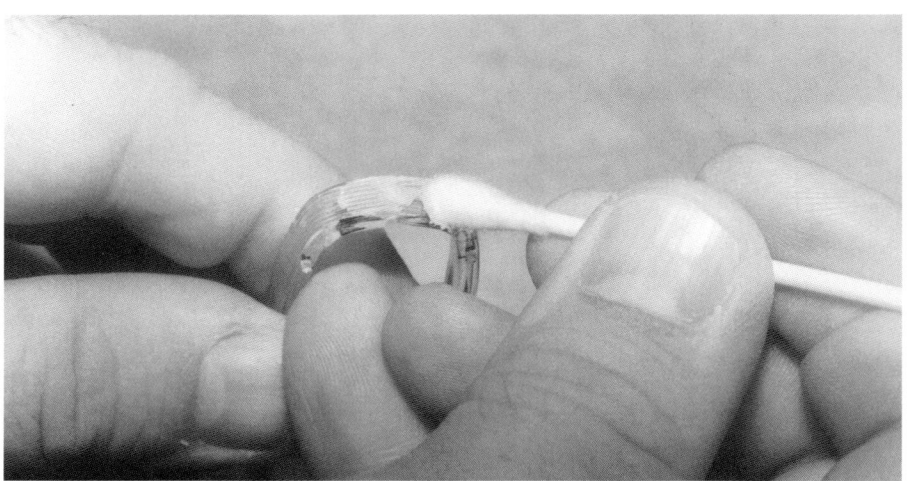

Fig 5-35. Using a polishing compound, remove the heaviest of the scratches.

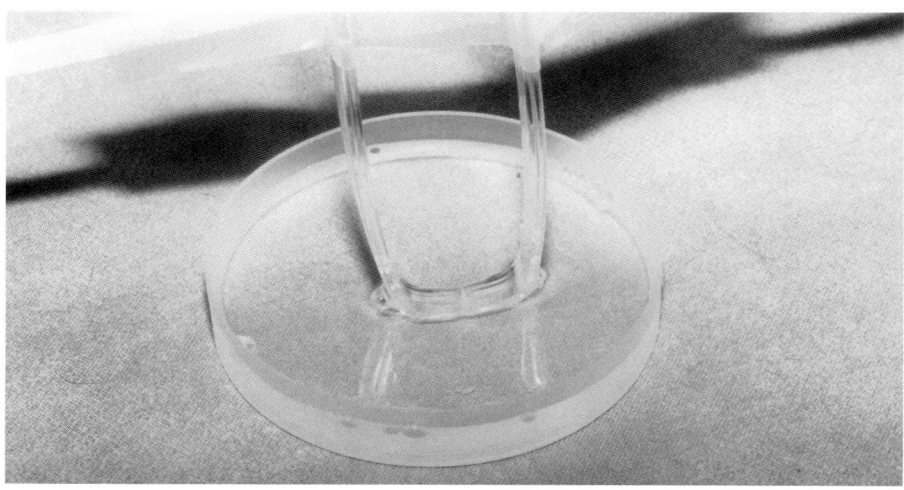

Fig 5-36. Dip the joint in acrylic floor wax to hide the remainder of the fine scratches.

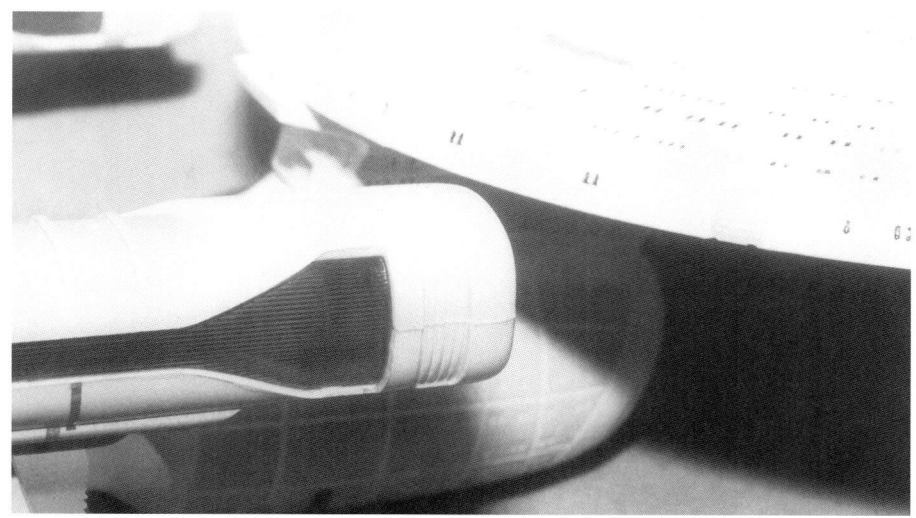

Fig 5-37. Some seams can't be handled until late in the assembly process.

Fig 5-38. Rock a curved blade along the groove to cut the drafting tape.

Fig 5-39. Mask the first and second colors before applying the final color.

After the second color has dried, apply a second mask to protect it from the application of the third panel color (fig. 5-39). After applying all the paint, don't remove any of the masking, since you will need it when you're repainting the joint between the top and bottom sections.

There is enough room behind the mounting tab to run the wire from the secondary hull into the primary hull (fig. 5-40). Glue the lower saucer to the secondary hull and fill and sand the joint. Prepare the wiring harness and attach it to the power wire (fig. 5-41). With a lot of room to work, you can attach the harness to the saucer with tape, epoxy, or even liberal amounts of white glue.

Glue the top portion of the primary hull into place. Mask around the joint as closely as possible, then apply filler. Remove the masking and let the filler dry. After sanding, touch up the joint with the base color. Now remove the masking and touch up the pattern as needed.

Step 6 – Cleaning the windows

After you have applied the last coat of paint to the model, it is time to "open" some windows to take advantage of all the wiring and lights. (After all, this is the reason you went to all this extra effort, isn't it?)

The simplest and most straightforward approach is the one given in the instructions. Using a pointed hobby knife, scrape the paint from the windows. I noted earlier that you had to leave holes in the interior paint mask. Now is the time to find those holes. Once the scraping is complete, touch up the paint one final time to take care of any overly zealous scraping.

Opening every window would look unnatural. To make some of the windows dark, use a mechanical pencil with a .05mm lead (fig. 5-42). The results are more realistic than using a black pen or paint. Do this before applying the gloss coats for the decals, but don't wait very long. Small bits of graphite dust will be left in each window you fill. As you handle the part, your fingers will pick up this powder and start to leave fingerprints all over the finish. Seal it in quickly.

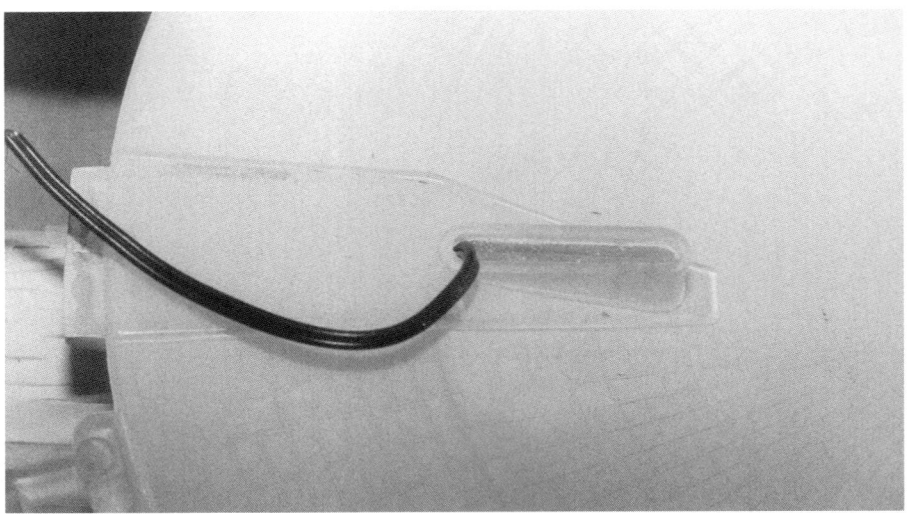

Fig 5-40. Run the power wire behind the mounting tab.

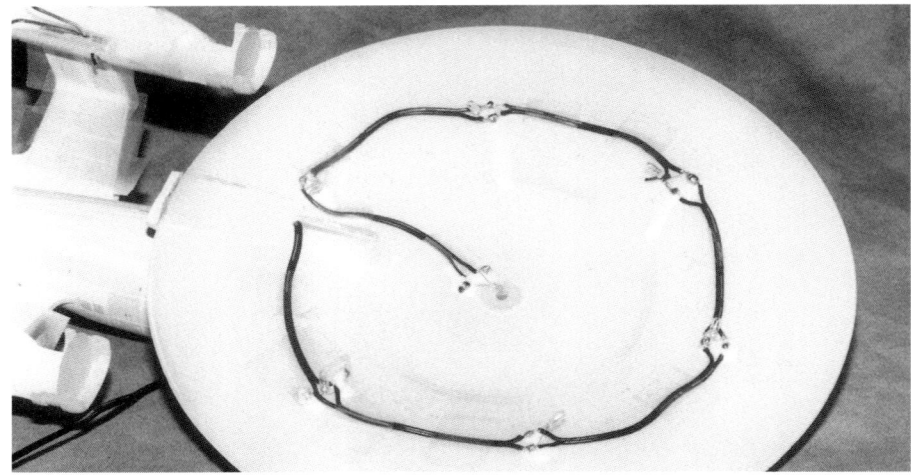

Fig 5-41. String the bulbs around the saucer to distribute the light evenly.

Fig 5-42. Add dark windows using a mechanical pencil.

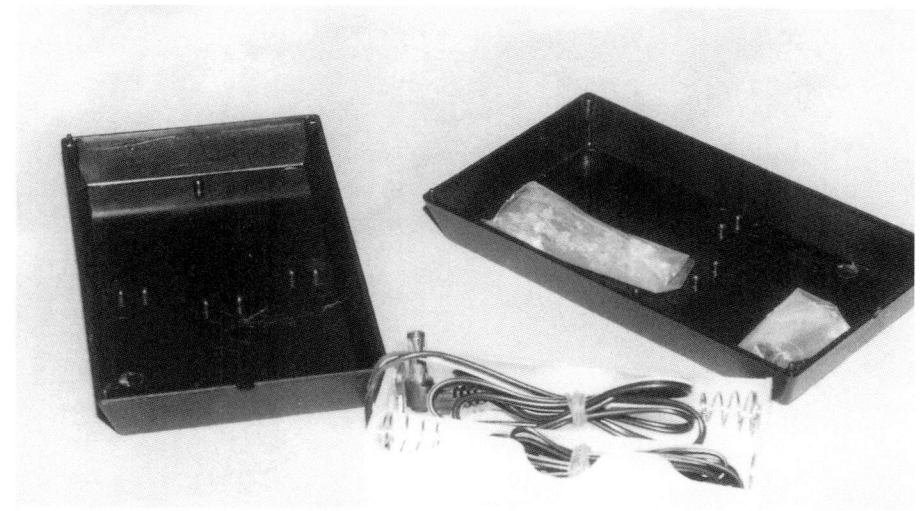

Fig 5-43. Power box for the kit

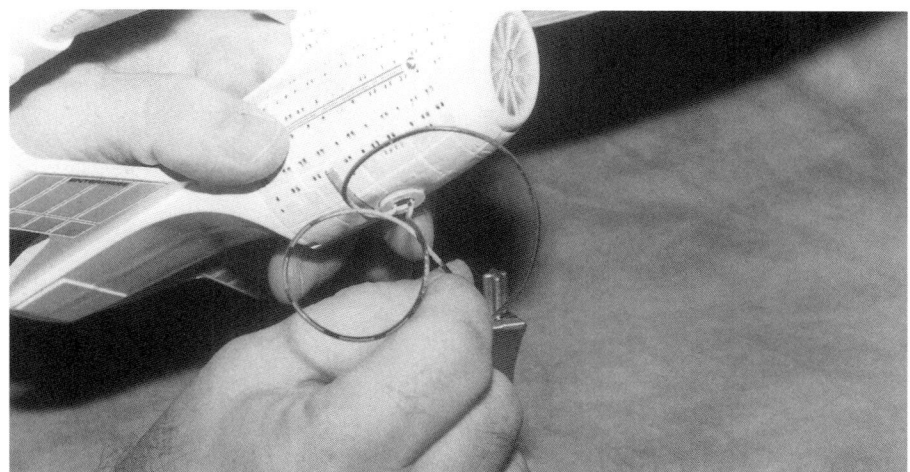

Fig 5-44. Feed the power wires into the neck of the base.

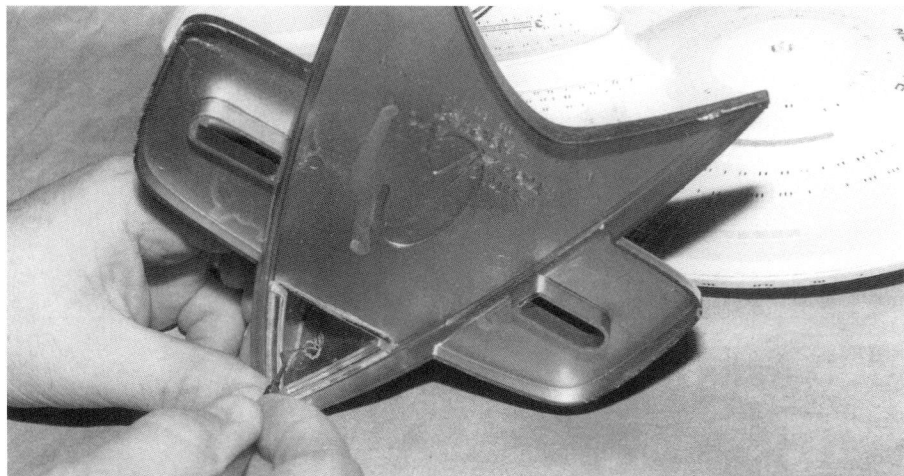

Fig 5-45. Pull the wires through a hole in the base so they can be hooked up to the power box.

Step 7 – Decals

Decaling this model is straightforward. Following the basic guidelines for gloss-coating the surface (see Appendix 2), apply the decals. Dull-coating (if desired) to seal the decals and provide a low-gloss finish is all that it takes to complete the kit.

Step 8 – Adding power

Follow the kit instructions for assembling the power box (fig. 5-43). Cut two small holes next to the mounting post in the neck of the base (fig. 5-44). Split the main power cord and feed one line through each hole. A larger hole is cut into the bottom of the base for the wires to exit (fig. 5-45).

6

Vacuum-formed Romulan Bird of Prey

The Romulan Bird of Prey from the Classic *Star Trek* TV series episode "Balance of Terror" had a brief but glorious contribution to the series history. A first-season episode, it combined many of the elements of social commentary that made the show famous. While many fans can replay the scenes with uncanny skill, the ship that played such a key roll had so little time in front of the camera that trying to replicate it accurately is rather difficult. Practically the only reference material comes from the episode itself (fig. 6-1).

I believe this lack of exposure is partly to blame for the small number of kits released of the ship. The Romulan Bird of Prey was released once by AMT in 1975 in a large-scale format, and it is also part of the 3-in-1 set that was covered in Chapter 1 of this book.

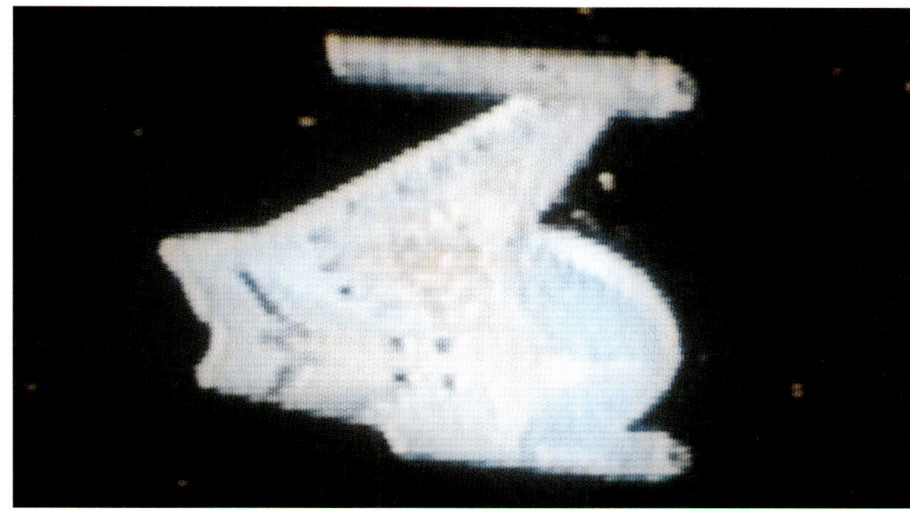

Fig. 6-1. When you're looking for reference data on the Bird of Prey, this fuzzy screen-capture shot from "Balance of Terror" is about as good as it gets.

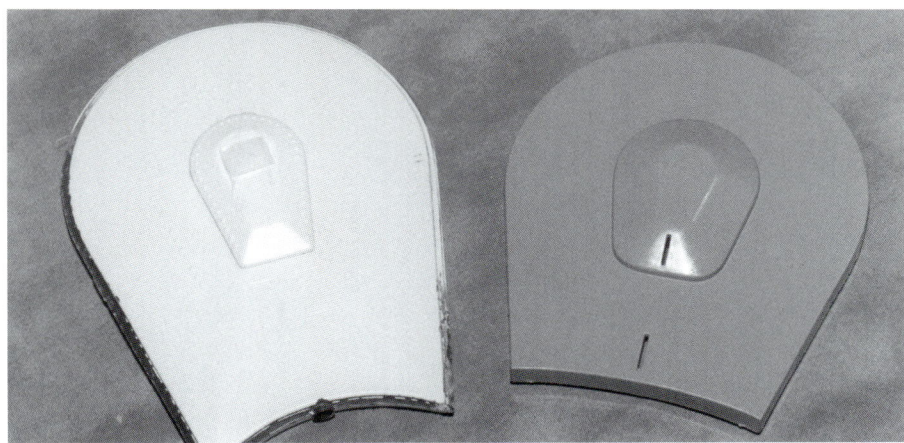

Fig. 6-2. This shot shows that the shape of the vacuum-formed kit is much more accurate than that of the AMT Romulan Bird of Prey released in 1975.

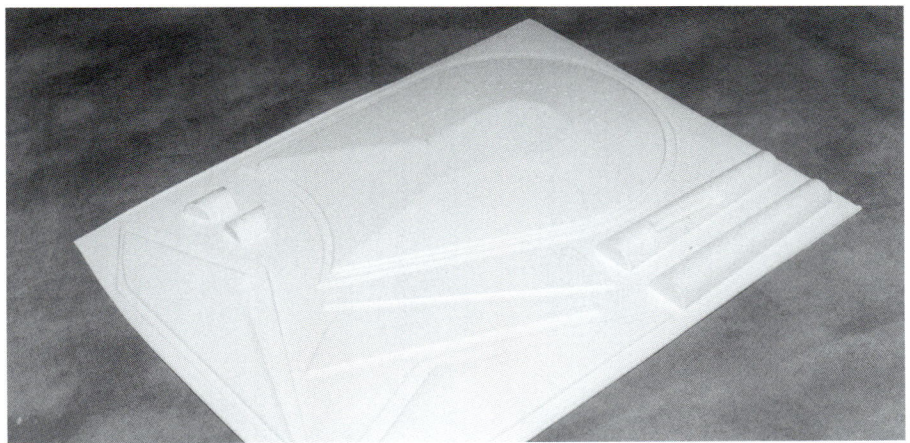

Fig. 6-3. Unlike injection-molded kits, parts for vacuum-formed kits come in sheets rather than on sprue.

This kit is vacuum-formed. Vacuum-formed kits have the same niche in the *Star Trek* universe that they have in other genres of modeling. They focus on subjects of "limited interest," meaning that the major modeling manufacturers believe the subject will not generate enough sales revenue to make the project feasible. This leaves production to the smaller "cottage industry," which usually works in the resin or vacuum-formed media that require less physical production overhead.

One of the more noteworthy examples of this is a vacuum-formed USS *Excelsior*. It was released many years before the AMT/Ertl kit and was supposedly in the same scale as the AMT/Ertl movie *Enterprise*, making it more than 30" long. Another example is the *Oberth*-class research vessel, which I have seen as both vacuum-formed and cast-resin kits. Notable examples of this class are the USS *Tsiolkovsky* from the Next Gen episode "The Naked Now," the USS *Grissom* from *Star Trek III: The Search for Spock*, and the cloaking device test ship USS *Pegasus* from the Next Gen episode "The Pegasus."

Both vacuum-formed and resin kits tend to arouse dread in the minds of modelers who feel they aren't up to the challenge of building this type of kit. While the steps are different, and some unique precautions are in order, you will find, to paraphrase the old commercial, that "parts is parts."

Vacuum-formed kits also have an unfounded reputation for being "unsophisticated," or inferior in quality. While this is sometimes true, it is also true that the vacuum-formed kit may be more accurate than any of the available injection-molded kits (fig. 6-2).

Manufacturers of vacuum-formed kits place a sheet of styrene or ABS plastic (fig. 6-3) over a female (recessed) mold. The mold has small air holes drilled in it. The sheet is heated until it is soft and then a vacuum is applied by sucking the air out of the spaces between the mold and sheet through the air holes. This draws the soft plastic into the mold and causes it to conform to the mold's shape (fig. 6-4).

The process results in the three characteristics of vacuum-formed kits—variable thickness, dimples, and the ridge of plastic represented by the thickness of the sheet.

Vacuum-formed kits are *usually* formed from sheets of .020 to .040-inch thick plastic (this kit is .040). This may seem strong as you flex the edge of the sheet, but you must remember that the larger, deeper parts will be much thinner because the plastic must stretch to cover a larger surface area. A reduction in thickness of 50 percent is not impossible.

Since you are going to subject the parts to a lot of stress during sanding and assembly, the first thing to do is reinforce the parts with plastic bracing. Glue strip styrene or even scrap sections of the kit sheet to the inside of the large parts (fig. 6-5).

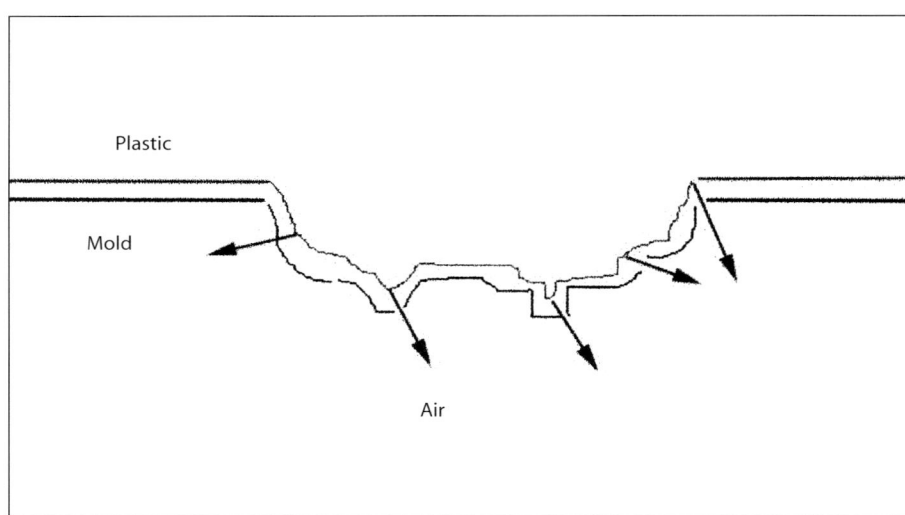

Fig. 6-4. The parts are formed by sucking the air from between the plastic sheet and the mold.

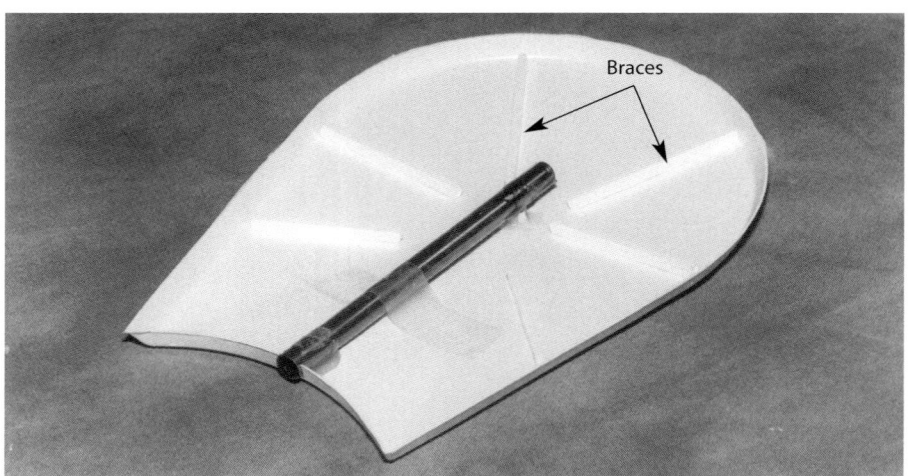

Fig. 6-5. Reinforcing the larger spans of the parts before sanding is a good idea. Run the pylon mount lengthwise through the model.

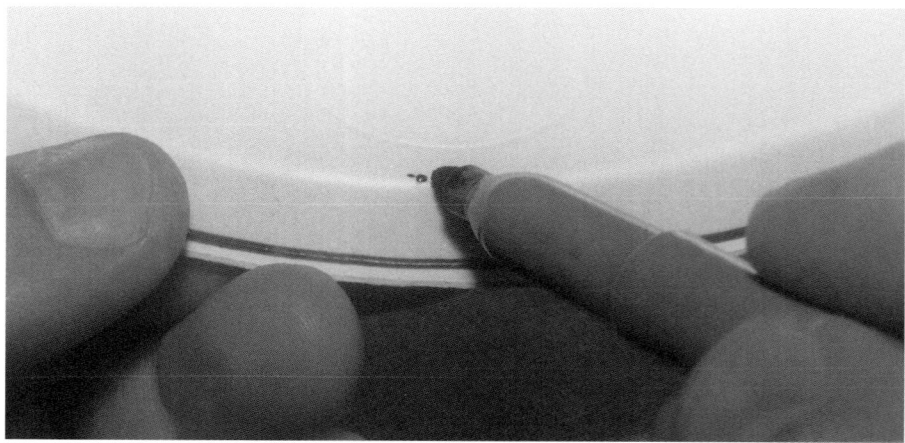

Fig. 6-6. Find the dimples and highlight them so you can sand them off.

Dimples form where the plastic is pulled into the vacuum air holes. They are usually rather small and can be found all around a part, but they are most frequent where corners are formed. Inspect each part for dimples by holding it up to a light and looking for the shadows they cast, or dragging a fingernail or knife edge along the surface until you feel a snag. Mark the dimples and the surrounding area with a marker and sand them away before painting the model (fig. 6-6). When there's no more ink, there's no more dimple.

Each part in the kit is the size of the recession in the mold. That is, if a part is 1" high, that's how deep the hole in the mold is for that part. If you cut the part out, you'll find that the height of the part is 1" plus the .020" to .040" that the thickness of the sheet adds. This may not sound like much, but it makes the difference between a model that looks good and fits properly and one that doesn't.

Gauging how much to sand and finding an efficient way to sand, are the real keys to making a vacuum-formed kit an easy exercise. Start by drawing an outline around each part with a dark-colored marker. Get as close as you can, leaving a mark on both the flat and the sides of the part itself (fig. 6-7).

As you are ready to assemble a set of parts, cut them out of the main sheet close to the edge, but leave a "halo" of the marked area around them (fig. 6-8).

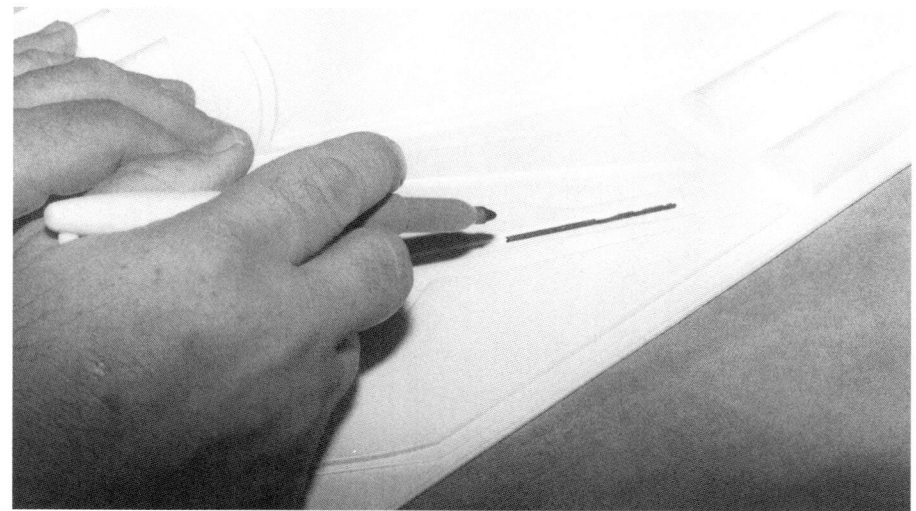

Fig. 6-7. Outline the parts with a felt-tipped pen.

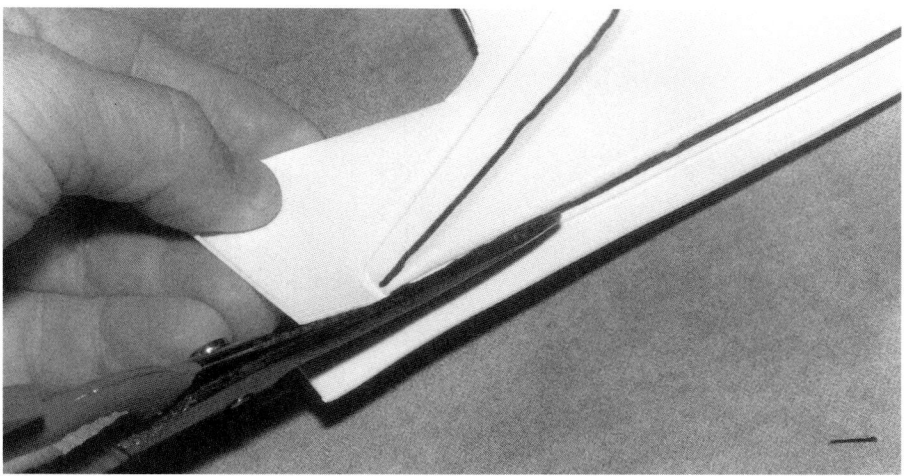

Fig. 6-8. Cut around the parts, leaving the outline.

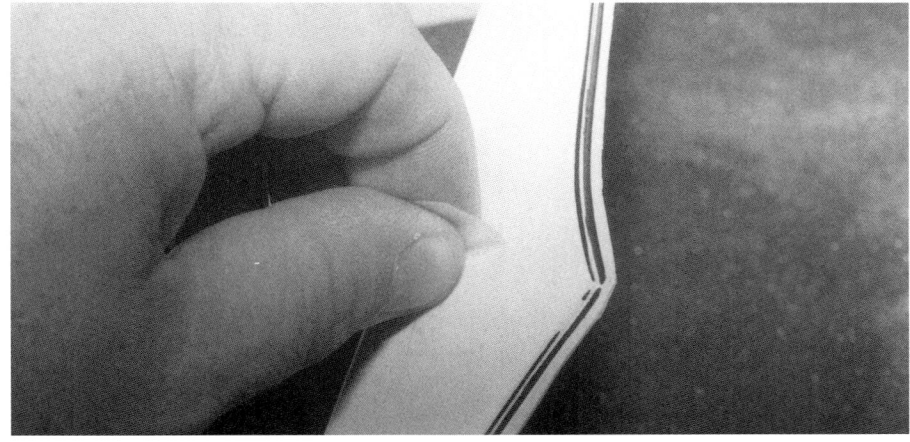

Fig. 6-9. You can make a "sanding handle" for hard-to-hold parts out of tape.

Tape a sheet of 320-grit sandpaper to a flat surface, preferably something that won't be scarred by overruns. Set the part on the sandpaper and begin to sand away the sheet area. If necessary, a handle of tape will make it easier to hold onto the part as you sand (figs. 6-9 and 6-10). As you get close to being finished, the plastic sheet will become thinner and thinner, and the "halo" will become translucent (fig. 6-11). Frequently check all around the part to see if unequal pressure is sanding some areas down faster than others, and adjust your sanding accordingly.

Once the "halo" is a mere wisp of plastic, you can gently tear it from the sides of the part and dress the edge with the sandpaper. Clean the sides of the part to remove any remaining flash, using the marker ink as a gauge (fig. 6-12). Sanding until the ink disappears indicates you have achieved a smooth edge. Remember my comment about parts thinning out. Don't oversand.

One advantage injection kits have over vacuum-formed kits is the alignment pins. Larger parts of a vacuum-formed kit are notorious for being partially out of alignment, no matter what. To solve this, glue strips of plastic (composed of the same material as the model itself) to the inside of a part that is part of a "clamshell" assembly (fig. 6-13). Select a strip wide enough to provide a good working surface but not so wide that it interferes with the other part you are trying to attach. The piece must also be thin enough to be fully flexible to the contours of the part.

Fig. 6-10. Sand the parts on a sheet of sandpaper taped to a large, flat surface.

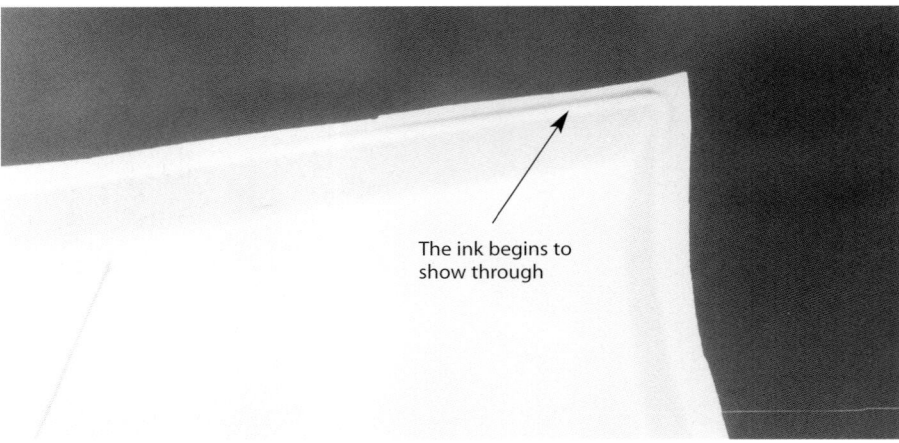

Fig. 6-11. As you sand away the thickness of the sheet, the "halo" becomes visible through the translucent plastic.

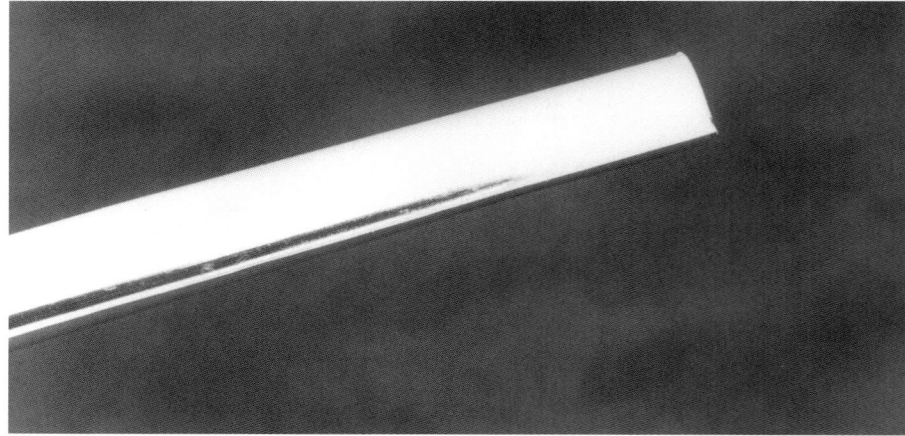

Fig. 6-12. The ink on the parts also helps to indicate when the excess plastic has been sanded off.

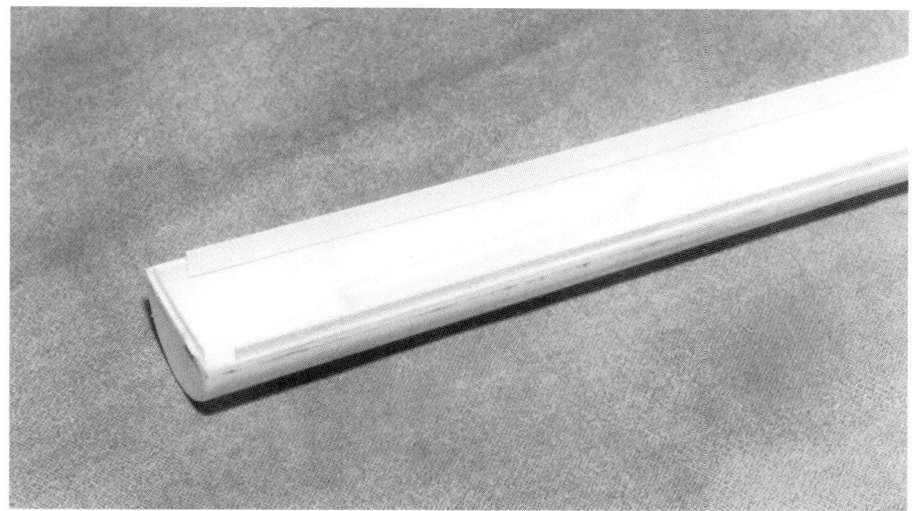

Fig. 6-13. Since there are no locating pins, laminate a strip of styrene to the inside of the parts to help with alignment.

Fig. 6-14. To improve the structural integrity, add a spar to each pylon. Use it to mark the location for gluing the pylon to the hull.

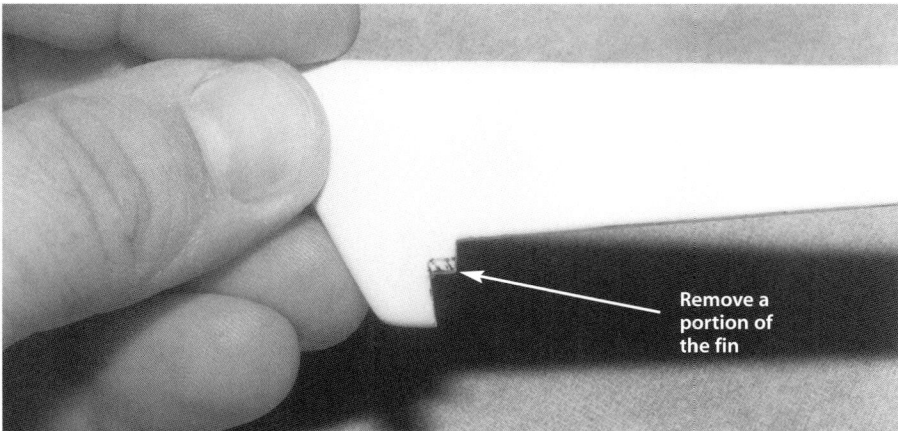

Fig. 6-15. Remove a bit from the fin to get it to fit.

Finally, you don't want the shim distorting the shape of the part. In most applications, a piece of .01" x .12" styrene will do the trick. As you apply the glue to the joint, it will flow along the butt joint and the shim, providing greater glued area and greater strength.

Where the pylon is attached to the main hull, add a square wing spar made from .020 x .040 styrene to help prevent flexing and to provide additional attachment points. Once the main hull and the pylons have dried, align the pylons to the side of the hull (fig. 6-14) and mark the proper location.

Before gluing together the hull halves, consider how you want to display the model. I wanted an "in-flight" display. Normally, I would have the tube for such a display extend from the bottom of the model, but in this case that would cause it to fall right in the middle of the markings. For this model, I located the tube along the central axis of the ship (refer back to fig. 6-5), extending out the back. The mounting rod will then make some type of 90-degree turn into a base.

After gluing and cleaning up the dorsal fin, test-fit it to the top of the hull. You should find that some portion of the sawtooth will have to be removed to get it to fit properly (fig. 6-15). In this case, I removed 2mm from the fin. Just as with an injection model, you'll have to fill and sand the joints. The quality of the kit will determine whether you are faced with just a few touch-ups or major reconstruction (figs. 6-16 and 6-17).

When a model like this requires a large amount of filler and, subsequently, sanding, it is usually better to wet-sand the model instead of dry-sanding it (fig. 6-18). Wet-sanding creates a slurry of the sanded filler. The water carries it away from the work area so it doesn't clog the sandpaper, making the sandpaper more effective. Wet-sanding also tends to leave fewer scratches in the finished product. When you have wet-sanded, it's easier to get a good paint and decal finish.

Fig. 6-16. Some of the joints are rather good.

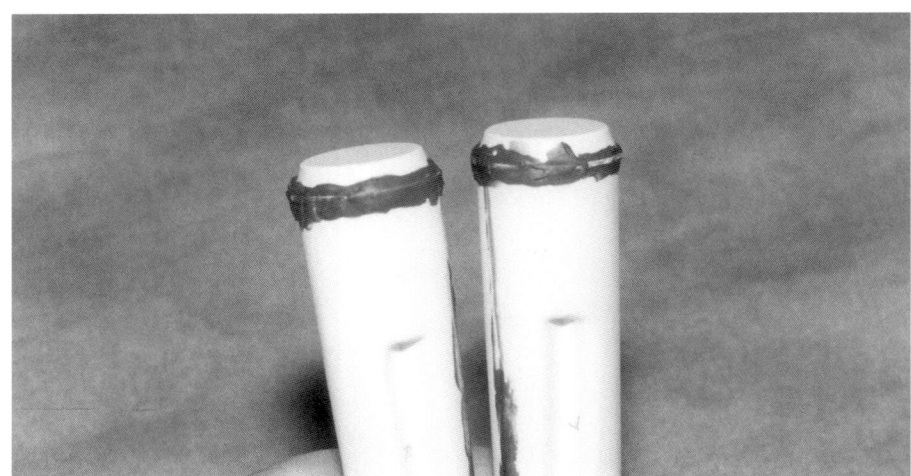

Fig. 6-17. Other joints need a lot of filling and sanding.

Fig. 6-18. Wet-sanding the filler provides the best result.

The plasma weapon is cast in resin. Cut the long pouring plugs from the rear of the part. The angle of the rear matches the angle of the hull so the barrels point straight ahead. Line up the part with the center axis of the hull and attach the weapon with super glue (fig. 6-19).

As you work with the nacelles, note that the recesses for attaching them to the pylons have a different shape. Mark them to help make sure that they are put on the correct side (fig. 6-20).

Spend time test-fitting the pylons to the nacelles. You should be able to get a snug fit along the top of the pylon when they are properly aligned (fig. 6-21). View this angle from the end so you can visualize the correct angle and can make sure you retain it after you apply the glue (fig. 6-22).

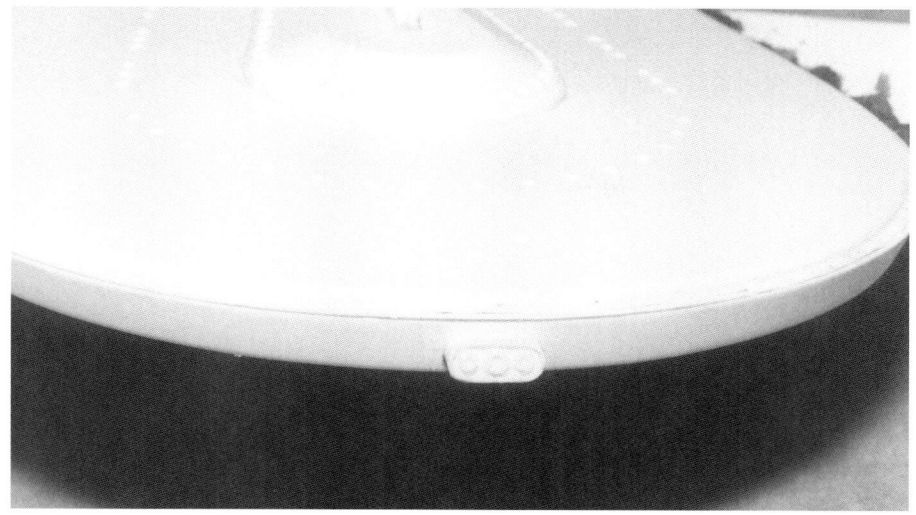

Fig. 6-19. Attach the plasma weapon to the front of the main hull.

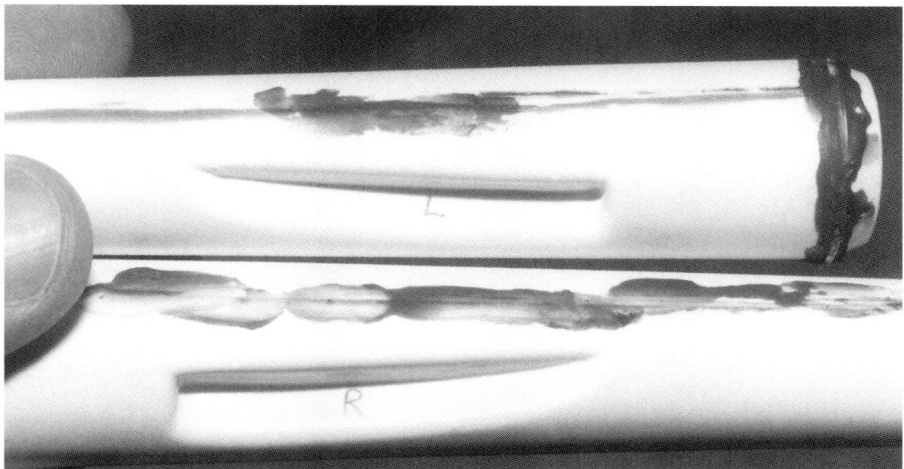

Fig. 6-20. Make sure the nacelles are placed on the correct side of the model.

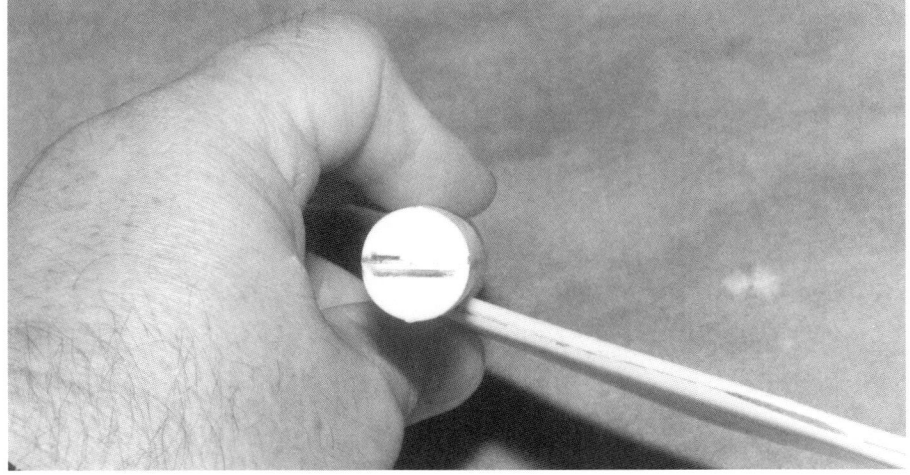

Fig. 6-21. Be sure to check the correct alignment of the nacelles before gluing. Before the glue sets, check it again and adjust if necessary.

The kit comes with both plastic exhaust tubes (which must be glued together) and turned aluminum tubes (fig. 6-23). Whichever option you choose, don't apply the exhaust tubes until you mount the nacelles and pylons. To make sure you have the diagonal correctly positioned, place a tick mark along the top (fig. 6-24). This will help you find the proper alignment when you glue it into place.

Perhaps the most critical seams on the model are the ones on the bottom where the pylons attach to the hull and the nacelles (fig. 6-25). They must be very smooth, blended seams, since the decals (the major selling point of the kit) must go over this area. Irregularities or inconsistencies in the two joints can spoil the work you've put into the rest of the model.

It's usually very difficult to see imperfections when you're working with dissimilar materials like plastic and filler. What you need is a consistent color and finish to check for problems. Apply a primer coat or a pre-coat of the hull color (fig. 6-26).

Once the paint is dry, check the joint for ridges, pits, and sanding marks. This should be done to the entire model, not just this particular joint.

Fig. 6-22. Check the alignment and angle of the pylons.

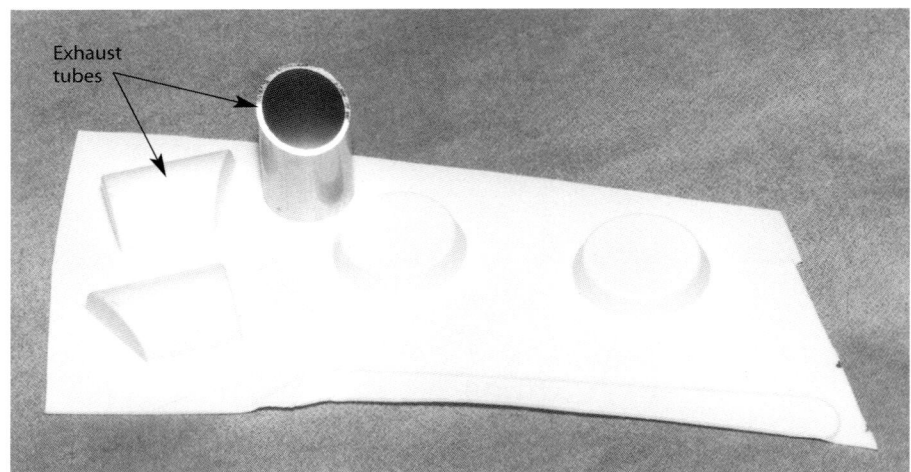

Fig. 6-23. You have the option of using the plastic parts or the turned aluminum parts for additional detailing.

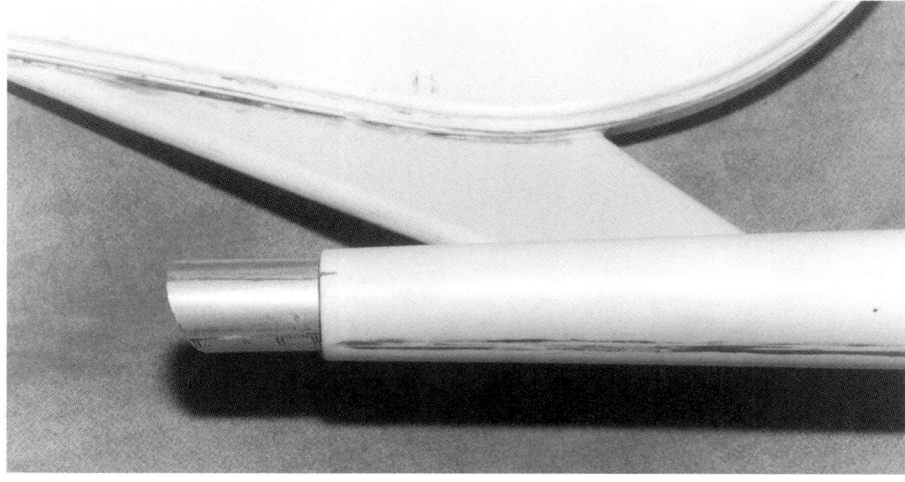

Fig. 6-24. Mark the top of the exhaust tube and the nacelle, then align the marks for gluing.

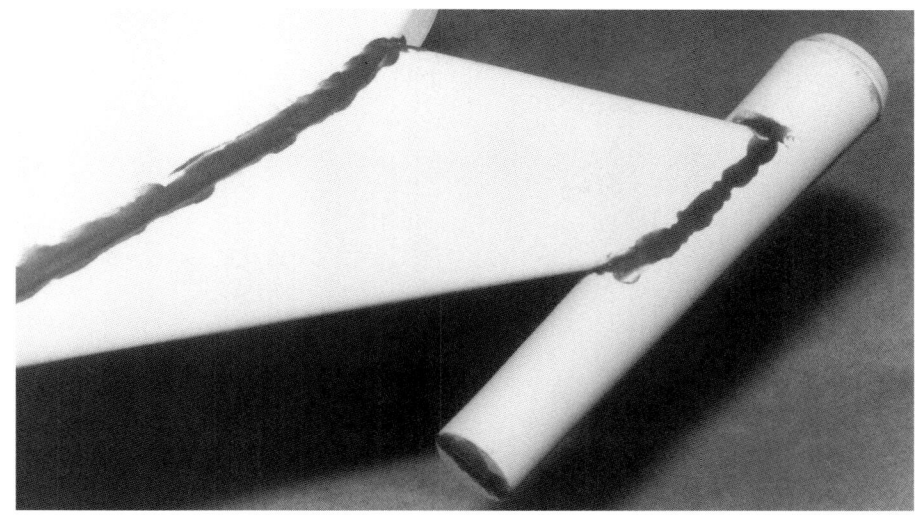

Fig. 6-25. Give a lot of attention to the pylon-hull seam so it will be a smooth surface for the decals.

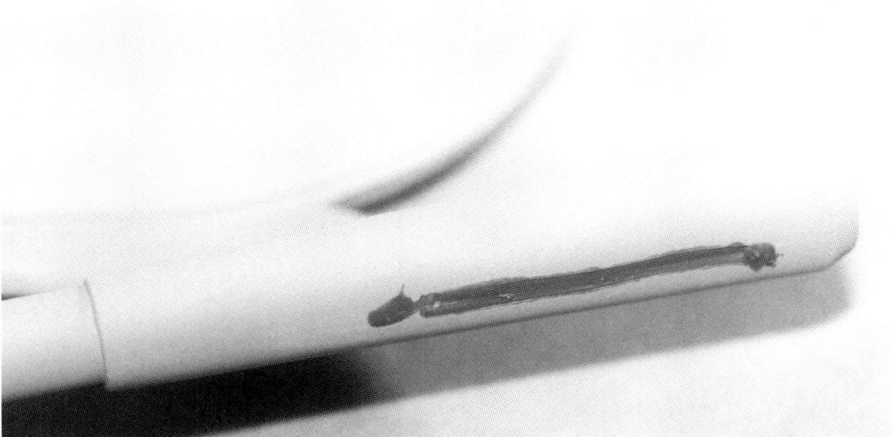

Fig. 6-26. Prime the model to find imperfections and repair them before applying the final coat of paint.

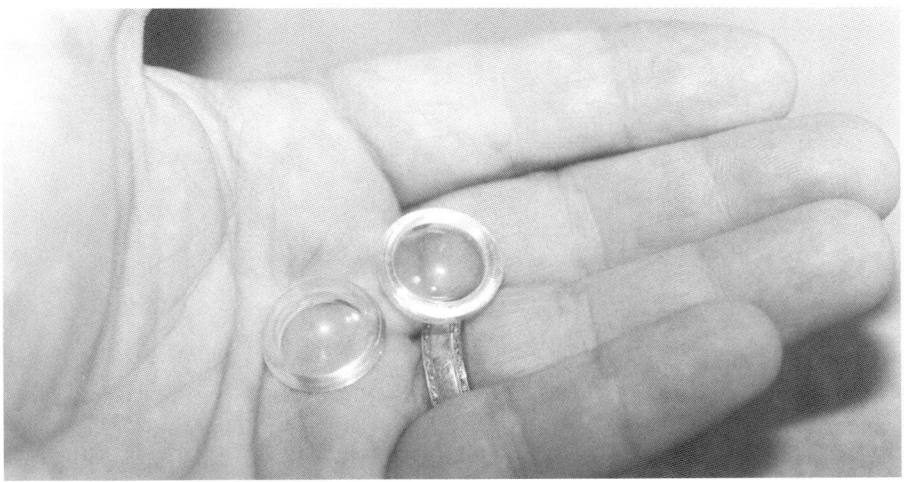

Fig. 6-27. The clear domes for the front of the nacelles are accurate, but they will benefit from a little extra attention.

The color (shade) of paint is open to speculation. This time I chose Testors ModelMaster Russian Topside Blue to provide a more intense blue than the Duck Egg Blue (FS 35622) used on the small-scale Bird of Prey in Chapter 1.

Another interesting difference between this kit and the AMT is at the front of the nacelles. This kit comes with two clear domes (fig. 6-27). My first inclination was to treat them the same as the bussard collectors on Federation starships. This would be incorrect since the Bird of Prey doesn't have warp drive, despite the similar shape of the nacelles. A frame-by-frame review of the episode "Balance of Terror" leads me to believe that the kit is correct. They are simply clear domes attached to the front of the nacelles.

I still felt the need to add a bit of pizzazz. First, disguise the thickness of the domes by painting the flat edge the hull color (fig. 6-28). Use a template to draw a circle slightly smaller than the diameter of the dome on the back of some metallic colored Mylar (fig. 6-29). Lay a bead of Micro Kristal Klear around the inside edge of the dome (fig. 6-30) and glue the Mylar to the back (fig. 6-31). Set these aside until you are ready to glue them onto the front of the nacelles after painting the rest of the model.

The decals for this kit are somewhat different from those found in mass-marketed kits. They have been applied to a solid sheet of clear decal film rather than having just a border of clear decal film surrounding the image (fig. 6-32). When cutting out the decals, trim as close as possible to the image without damaging it (fig. 6-33).

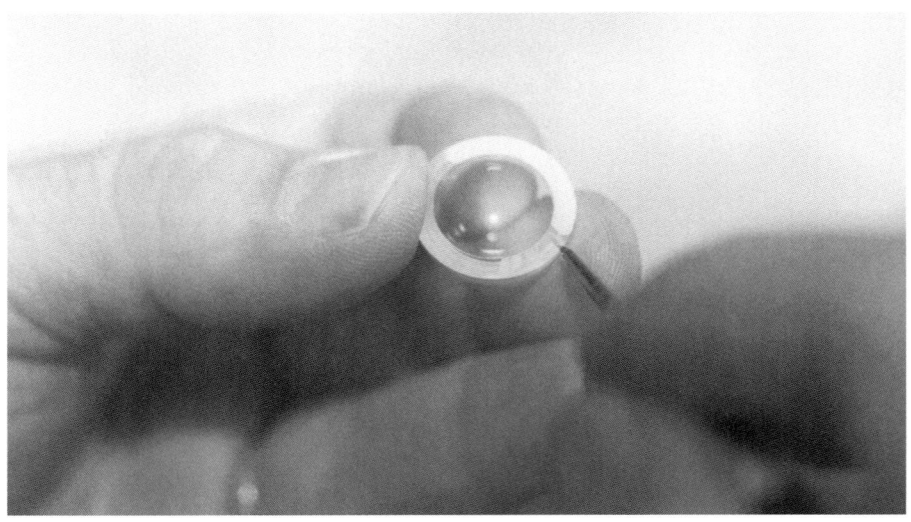

Fig. 6-28. Paint the back edge of the dome to make the thickness less noticeable.

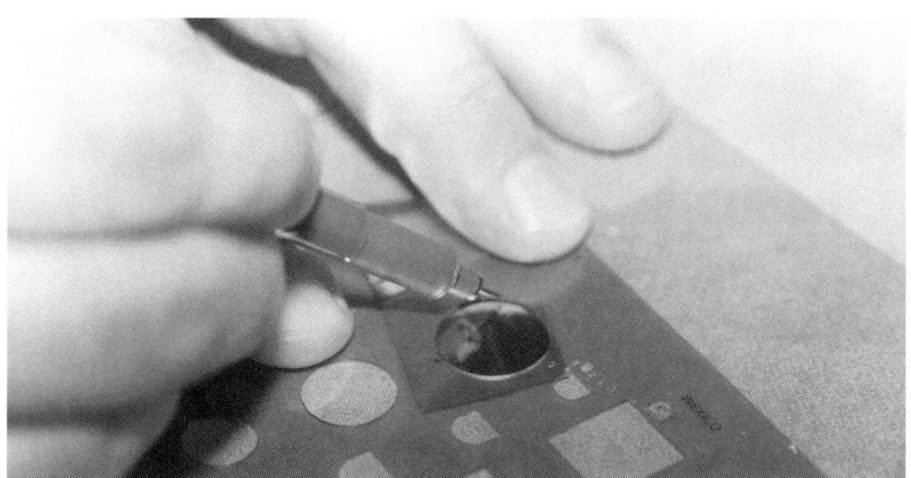

Fig. 6-29. Trace the pattern on the back of the Mylar.

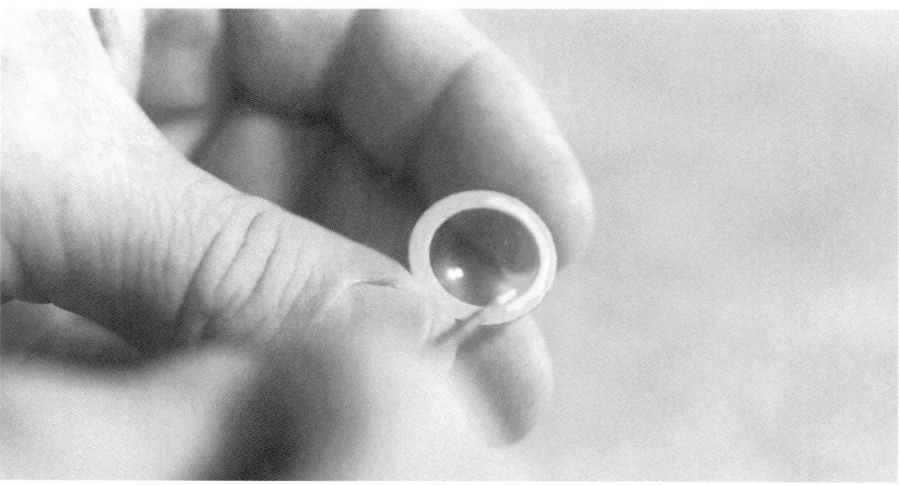

Fig. 6-30. Lay a bead of Kristal Klear along the edge to glue the Mylar to the dome.

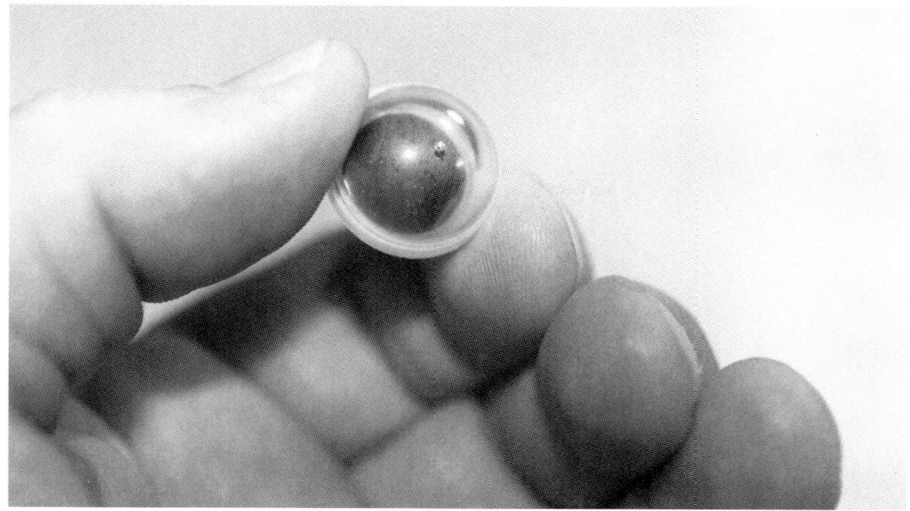

Fig. 6-31. I chose blue to add some color behind the dome to liven up the model.

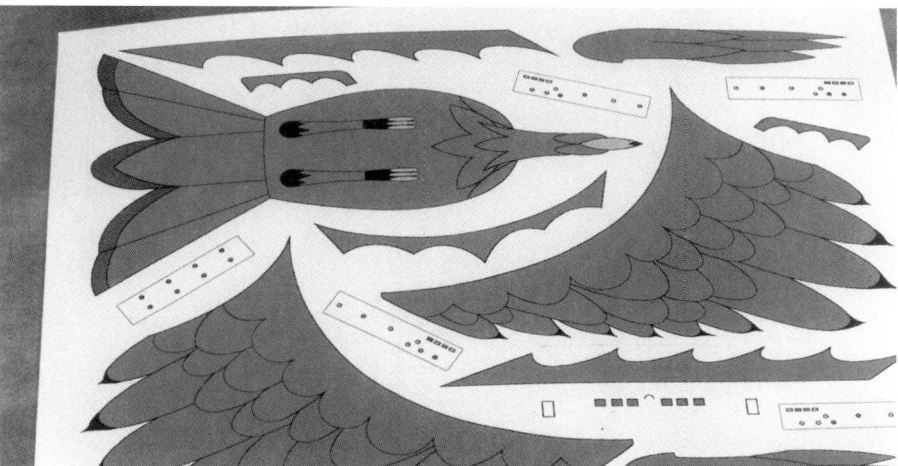

Fig. 6-32. The decals are printed on a solid sheet of decal film.

Fig. 6-33. Trim as close to the image as possible to reduce the possibility of silvering.

Just as the kit is a limited production run, so are the decals. You probably won't be able to get replacements if you mess up, so there is no margin for error. Before applying any of the decals, test a scrap piece with the decal-setting solution you normally use to see if you get any unexpected reactions. You may find that the decals work just as well using no solution.

The bird on the bottom of the model comes in five pieces. To maintain proper alignment on the multi-part decal, start with the body of the bird. Work your way outward in pairs with each section of the wings to ensure that they remain symmetrical (fig. 6-34).

The sheet includes several decals designed to add windows and other details around the edge of the main hull. Since I had no pictures of the studio model and the TV images are fuzzy, I chose to apply them, since their existence makes sense (fig. 6-35).

Each set of windows is surrounded by a black border (fig. 6-36). This doesn't seem to belong on the model, but more important, the sets are rectangular rather than curved. Since they go on a curved, tapered surface, the difference in shape causes the border to crawl across the surface away from the shorter edge (fig. 6-37). Trim the end of the decal having the most empty space to a wedge shape, allowing it to lie flat and not overhang the edge (fig. 6-38).

Fig. 6-34. To maintain balance when dealing with a multi-part decal, always work from the middle out.

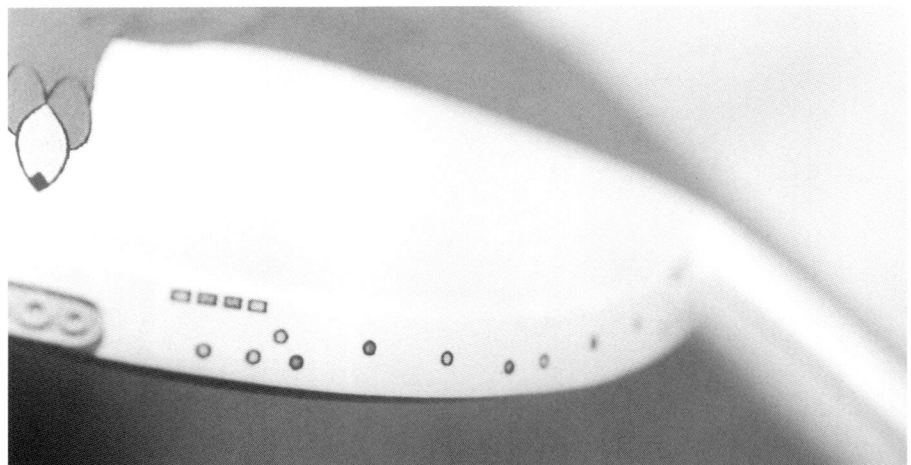

Fig. 6-35. While they are not clearly visible on television, it makes sense that there are portholes on the model.

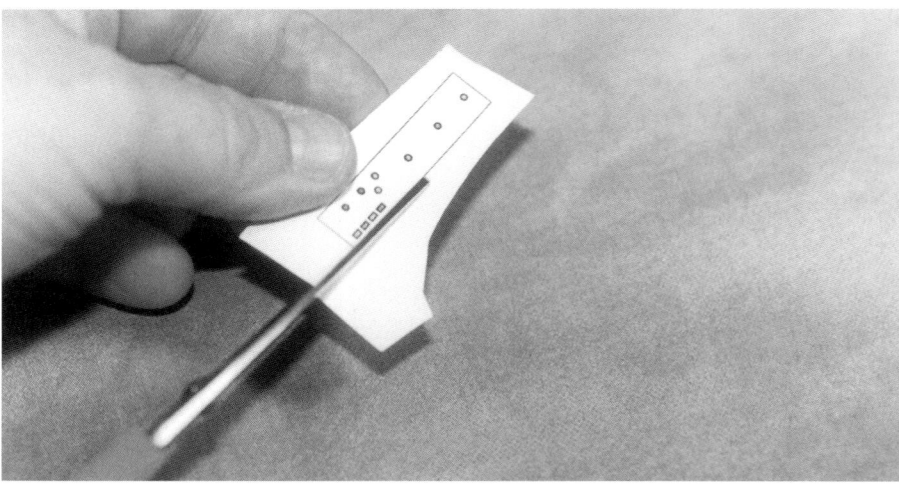

Fig. 6-36. Remove the border from around the porthole decals.

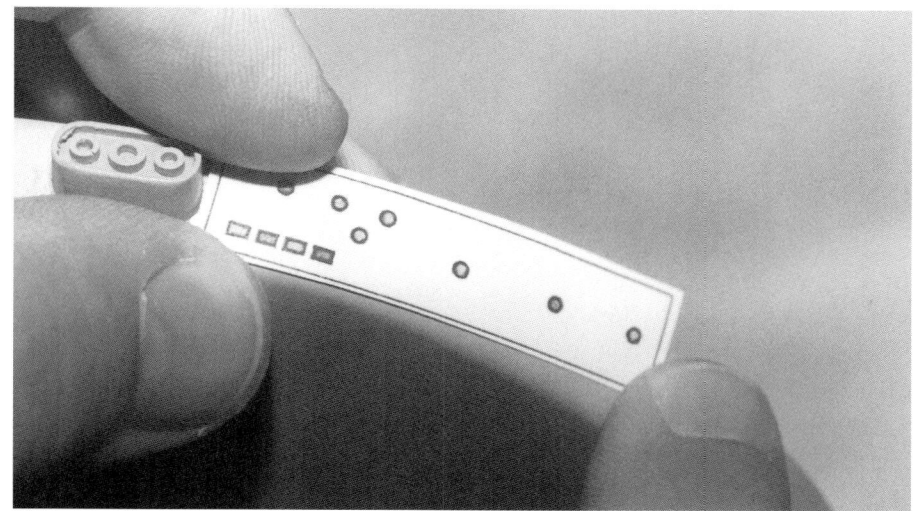

Fig. 6-37. Straight decals don't get along with curved surfaces.

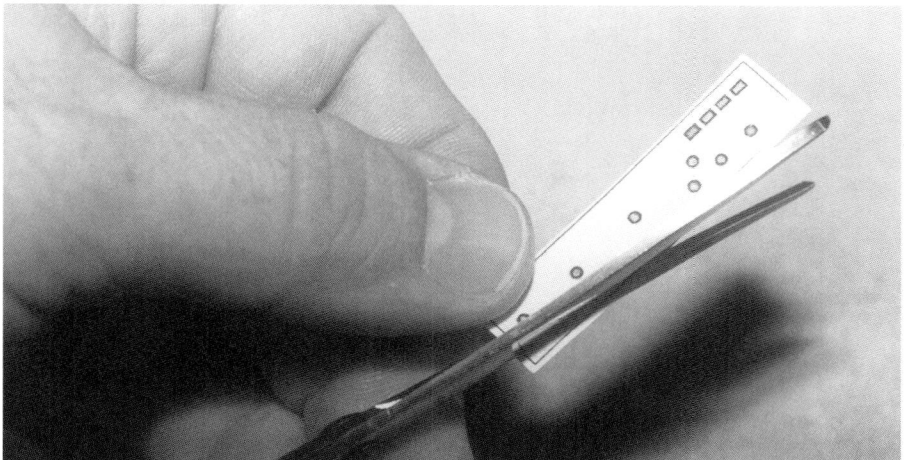

Fig. 6-38. On the end with the most blank space, trim the decal into a point.

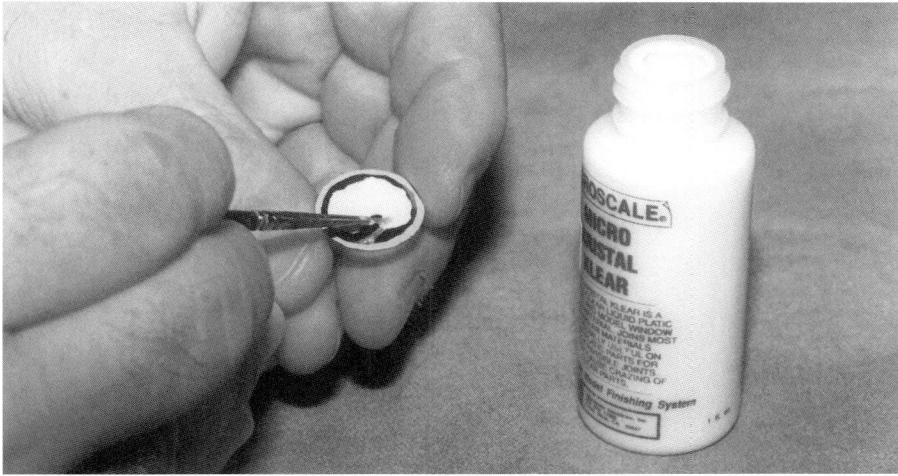

Fig. 6-39. Use Kristal Klear as an adhesive for attaching the nacelle domes.

Finally, there is a series of decal scallops designed for the upper pylons and the main fin. Even though the manufacturer's research had been right to this point, they just didn't look as though they belonged. I decided to leave them off, but you might choose to do otherwise.

The final steps are to apply a flat finish to the model and then attach the domes to the front of the engines. The Kristal Klear is again a good choice for the adhesive (figs. 6-39 and 6-40).

Most modelers who work in both media feel that vacuum-formed models aren't that much more difficult to build than the injected kind. They require you to provide some of the features normally found in injected kits, but nothing extremely difficult. The biggest drawback to these kits is that they have limited distribution and very uncertain production runs. So if you see a kit that makes you think, "Someday . . . ," you should probably grab it. The source may dry up, leaving kits like these the only game in town.

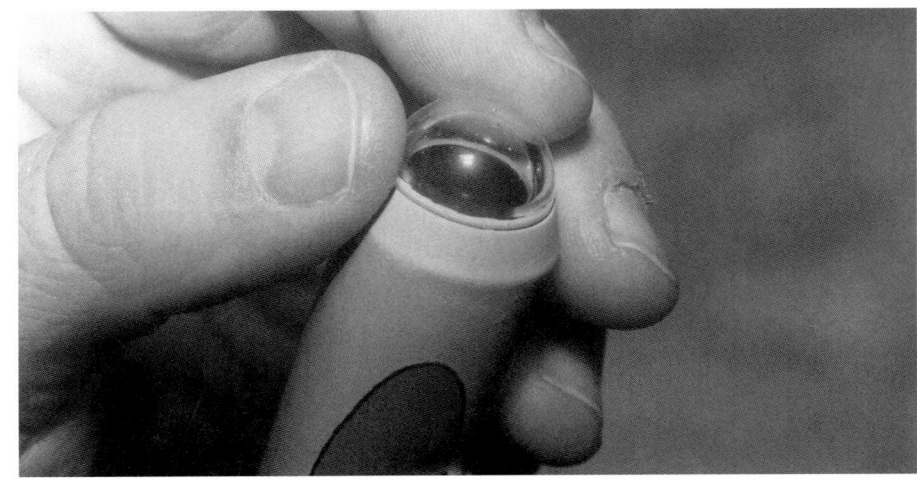

Fig. 6-40. Gently twist the dome as you seat it to even out any lumps in the Kristal Klear.

7

The Cardassian Galor-Class Cruiser

Star Trek: Deep Space Nine gave us a look at new villains. Introduced in the *Star Trek: The Next Generation* episode "The Wounded," the Cardassians came into their own on DS9 and proved to be formidable adversaries. The *Galor*-class cruiser is the primary vessel for the Cardassian Empire.

Basic assembly

The *Galor* is a very simple kit, consisting of 40 parts: 18 of them are made of gray plastic and 22 are clear.

Part number 2 is the underside of the main hull. Even though it isn't specifically mentioned in the instructions, you will need to remove the braces (fig. 7-1) molded into the part before attaching the sides of the hull (parts 7 and 8).

The fit of the top and bottom parts of the main hull (parts 1 and 2) isn't very good, particularly at the aft end of the main hull (fig. 7-2), so be prepared to spend some time filling and sanding this area.

Assemble the parts before painting, but leave the upper bridge area (parts 3 and 4) separate until you've finished painting. There are details on the underside of the bridge as well as details on the top of the main hull that will be hard to paint and mask if everything is attached (fig. 7-3).

After assembling the main hull, you will find a gap between the sides and the underside of the "fan." To fill this area, lay down a strip of cellophane tape about 3mm from the gap. Apply the filler, working it into the gap (fig. 7-4). After it begins to dry (when the skin of the putty is dry to the touch), pull up the tape, removing the excess filler. Don't wait too long. The filler will attack the tape and cause it to bond to the model if it is left too long.

Sand the area carefully, avoiding as much of the surrounding detail as possible. As a final touch, remove the last bit of filler by scraping a curved no. 1 X-acto blade along the joint (fig. 7-5). This focuses the cleaning on as small an area as possible and makes it easier to achieve a crisp, 90-degree angle as well.

Painting

At first glance, the *Galor* seems a monochromatic ship. It does use two shades of tan, ModelMaster Radome Tan (FS 33637) and Sand (FS 33613), but these are very close to the same shade. Larger detail panels are either black or a yellowish-brown similar to RAF Middlestone with an FS number of 30266.

Fig. 7-1. Braces are included in the kit to protect the bottom plate from being broken off. Remove them before attempting to attach the sides of the hull.

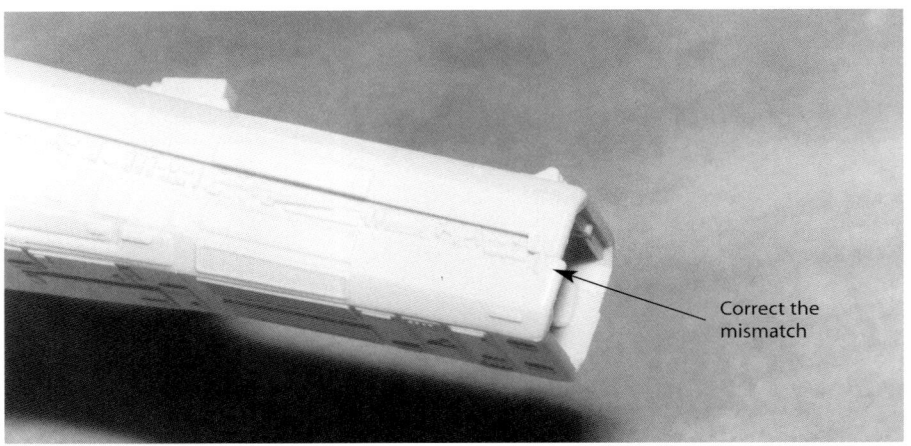

Fig. 7-2. You'll have to address some fit problems.

Fig. 7-3. Unless you leave off the bridge, painting and detailing the areas under the overhang will be difficult.

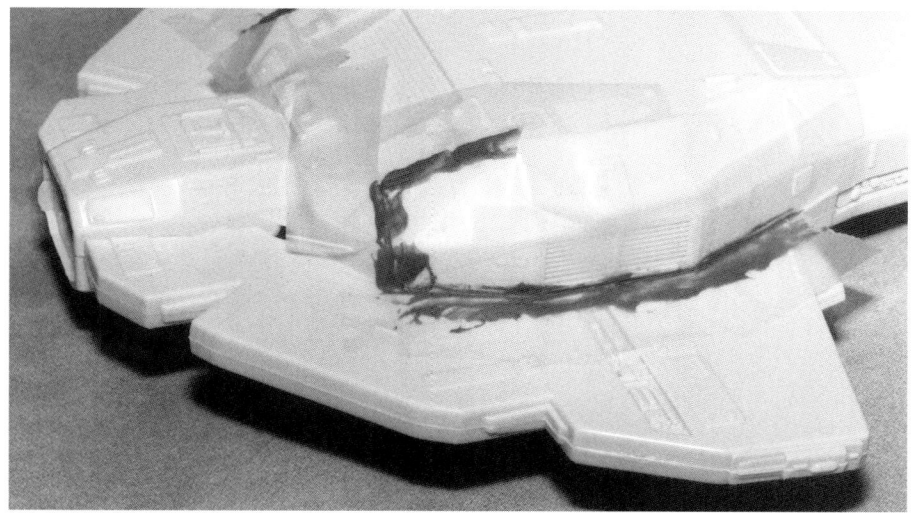

Fig. 7-4. Large areas of the model need filling. When working with right-angle areas, use cellophane tape to protect the surrounding sections as you push the filler into the corner.

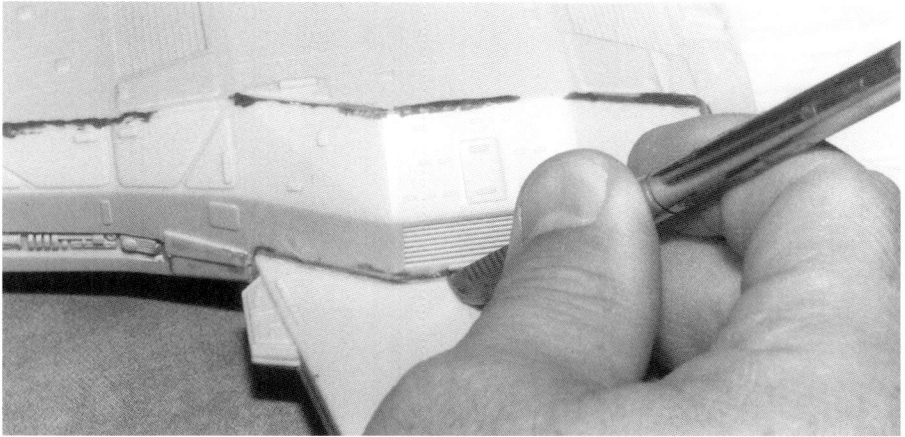

Fig. 7-5. Use a no. 1 X-acto blade to clean up the joint and produce a square edge.

Fig. 7-6. Apply the base coat of paint. Airbrushing is preferable, but you can apply the base coat by hand if you thin it properly.

To dress up the model, spend a lot of time highlighting panels and shading to create extra depth and character. By using a combination of paint types and various ratios of paint and thinner, you can accomplish this with ease. Use a combination of acrylic and enamel paint—a large variety of both is now available.

After painting the ship an overall Radome Tan, paint the sensor panels the Middlestone color using the acrylic paint. Thin it with isopropyl alcohol, 50/50 paint/alcohol for airbrushing or 80/20 for hand-brushing. I chose to hand-brush this time (fig. 7-6).

The thinned mixture will settle away from edges and high points, creating light edges and mimicking the effect of drybrushing. Because it and the alcohol are chemically different from the enamel base color, the high percentage of "thinner" won't dissolve the base paint. As a side note, don't miss the small sensor panels on each side of the upper superstructure. These aren't marked with a color key "J" on the painting instructions.

An easy way to apply a wash is to go back to an enamel paint (or acrylic if the base color is enamel) that is several shades darker. A quick rule of thumb is to choose a color that has an FS number at least 200 points darker than the base color and is in the same family (in this case, brown on brown).

The paint for the wash must be unused or must have settled for several days so there is very little pigment in the carrier. Test the paint on a card. If it is too thin, give the bottle a *single* shake and retest it until you get the intensity you want.

Apply a heavy coat of the wash to the detail (fig. 7-7). It will seem darker and over-shaded at first because of the carrier, so wait to evaluate the effect until it has dried. Notice the difference between a washed panel and one with just the base coat (fig. 7-8).

Before highlighting the panel lines, give the model an overcoat of clear gloss and apply the decals.

You want a glossy finish for two reasons. First, a flat surface is rough. The wash will have a tendency to spread out from the panel lines around the grain of the paint. On a glossy surface, it will tend to follow the panel line, keeping the pigment concentrated where you want it. Second, you want the weathering to affect the markings as well as the hull. As I noted in Appendix 2, decals must be applied over gloss to prevent silvering.

If these kits had recessed panel lines, you would be ready to start weathering. The panel line should be the darkest part of the detailed area. Without an additional step raised panel lines will end up being lighter than the surrounding area. Drag a new curved blade, such as a no. 1, along the panel line. Make sure the only point where the blade touches the model is the crown of the panel line (fig. 7-9). Gently scrape off the paint until the panel line becomes darker than the surrounding surface (fig. 7-10).

Using the same paint and technique as when you applied the wash for the sensors, carefully draw a line of paint along the panel line (fig. 7-11). You can do fairly large areas at one time, but be patient: don't try to do the entire model at once. The wash will take a few hours to set completely. Be careful not to smudge the finish by holding the model in an area that's still wet while you're trying to wash another area.

Fig. 7-7. A very thin wash of the darker brown settles in the crevices of the detail and causes it to jump out.

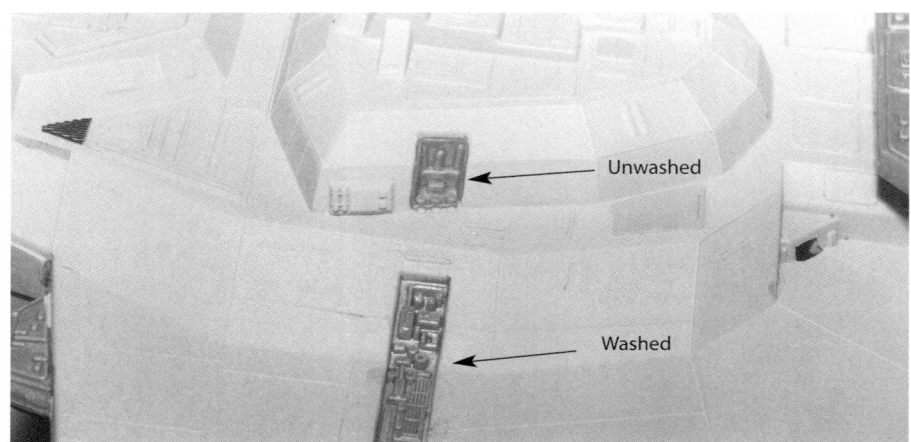

Fig. 7-8. Notice the difference between the sensors with the wash (bottom) and without (top).

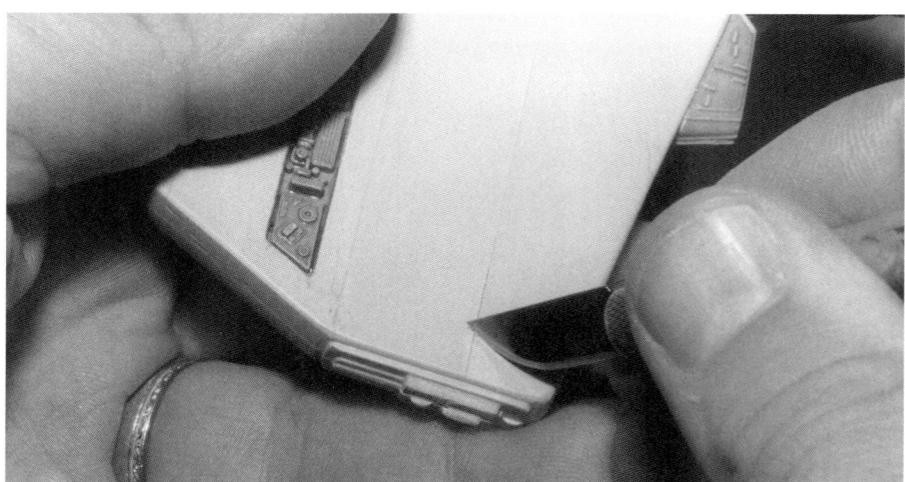

Fig. 7-9. Again using the no. 1 blade, scrape the panel lines to knock off the light paint and expose the darker color underneath.

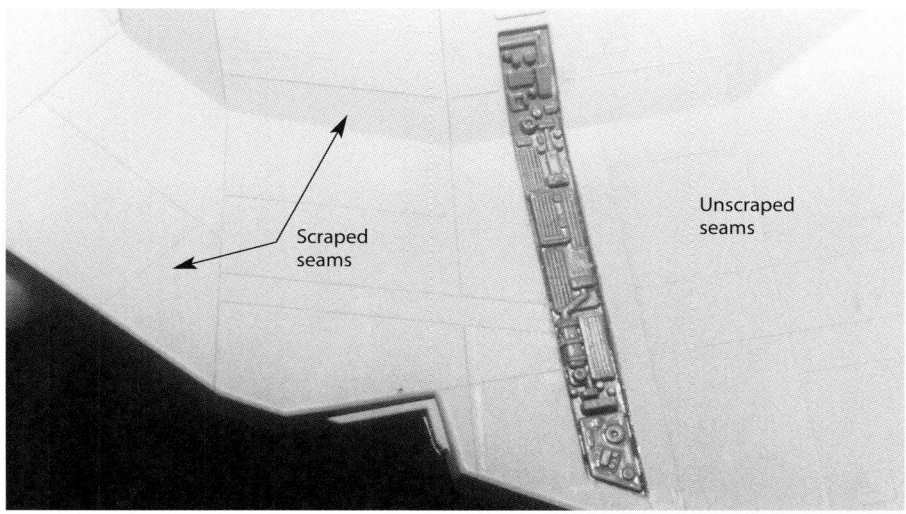

Fig. 7-10. To the left of the sensor array are panels that have been scraped. Notice how they are more distinct.

Fig. 7-11. On the left are panel lines after the application of the wash.

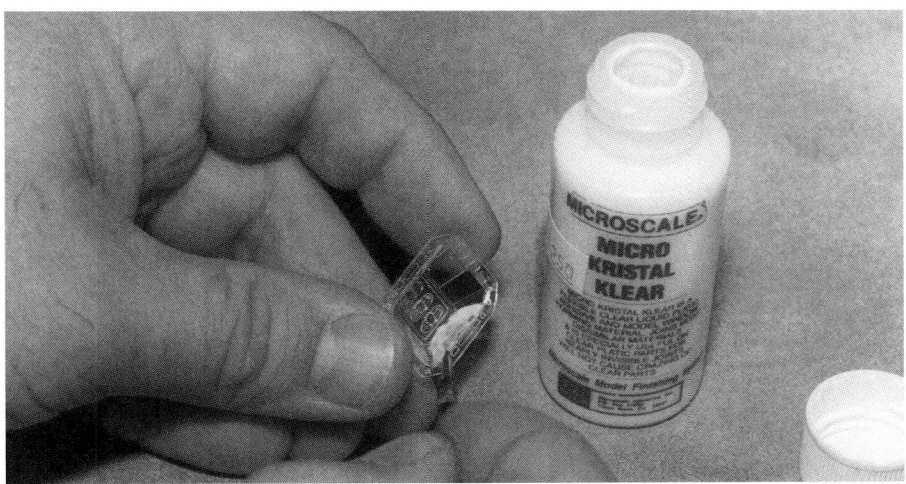

Fig. 7-12. You can use Kristal Klear as a contact adhesive. Since it dries clear, it won't distort the metallic finish of the piece.

You can highlight raised plates by tracing around the edges with the wash. Puddle the wash in the crevices of irregular shapes to intensify the shadows in the area.

Clear parts

Before adding any of the clear parts, spray the entire model with a flat clear overcoat and let it dry completely.

Part 103 is the main weapons array. Different references will lead you to different ways to paint it. According to the instructions and the box art, you are to paint the panels copper and the raised ribs a transparent blue. They have an internal glow with an effect similar to the Federation warp engines. However, in several episodes the panels are areas with a yellow/copper glow and the ribs are dark, probably the same shade as the main hull. I chose to go with the latter coloring.

I wanted to create a metallic effect for the main panels. At the arts and craft store I picked up some metalized Mylar sheets used for party decoration and some metallic gift-wrapping ribbon. Between the two, you should be able to find all the colors you need to make these kits.

Trim the ribbon to the proper shape so it will fit along a flat section of the part. Spread a thin film of Kristal Klear on the inside of the part, then press the ribbon into place (fig. 7-12). The Kristal Klear will form a milky film sandwiched between the parts. Work it around until you get uniform coverage. Since there is little exposure to the air, it will take quite a while for the Kristal Klear to dry, but it will dry clear, and the metallic effect will be retained (fig. 7-13).

Paint the outside ribs separately, or if you want, attach part 103 and mask it for painting along with the main hull.

Parts 105 and 106 are the impulse engines. The kit instructions tell you to paint them white and the raised ribs black. Again using the episodes as a guide, I back-painted the engines orange. To make them appear lit, paint the recessed area behind the parts chrome silver before cementing them in place.

According to the instructions, these parts are to be attached while the hull is being assembled in steps 4 and 6. There is a slight tongue-in-groove for the part to fit into. To protect them from overspray, I recommend leaving the parts off until the model has been painted. Drag a knife blade along the top and bottom edges to trim them slightly, and they will pop into place without any problem.

Clear parts 101 and 102 are tinted orange. Normally, I would recommend back-painting these parts to avoid any surface texture problems, but these parts have very thick faces. Back-painting leaves a prominent edge that remains clear (fig. 7-14). Carefully paint the pyramid-shaped formation lights on the underside as shown in the instructions. The lights are tiny and there aren't any extras in the kit.

Fig. 7-13. After installing all of the plates, paint the ribs and install the part.

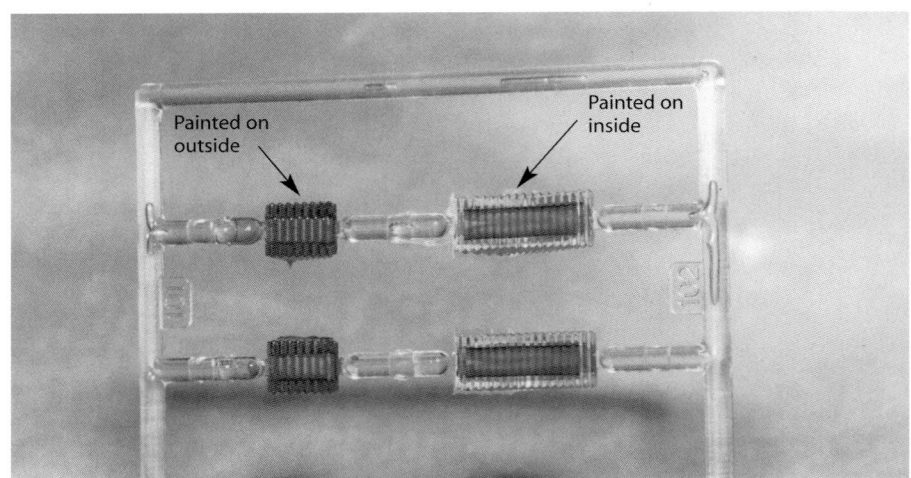

Fig. 7-14. On the right, the thickness of the piece creates a clear "halo" when painted only on the inside.

Photo Galaxy

Romulan *D'deridex*-class Warbird from AMT/ERTL. Model by Jerry Taylor

USS *Enterprise* NCC-1701-A from AMT/Ertl. Model by Scott Wilson

Above: USS *Enterprise* NCC-1701-C from AMT/Ertl. Model by Rick Jackson

Top right: Runabout *Rio Grande* from AMT/Ertl. Model by Rick Jackson

Bottom right: USS *Enterprise* NCC-1701-D from AMT/Ertl. Model by Rick Jackson

Above: USS *Defiant* from AMT/Ertl. Model by Rick Jackson

Top right: USS *Enterprise* NCC-1701-E from AMT/Ertl. Model by Rick Jackson

Bottom right: USS *Voyager* from Revell/Monogram. Model by Rick Jackson

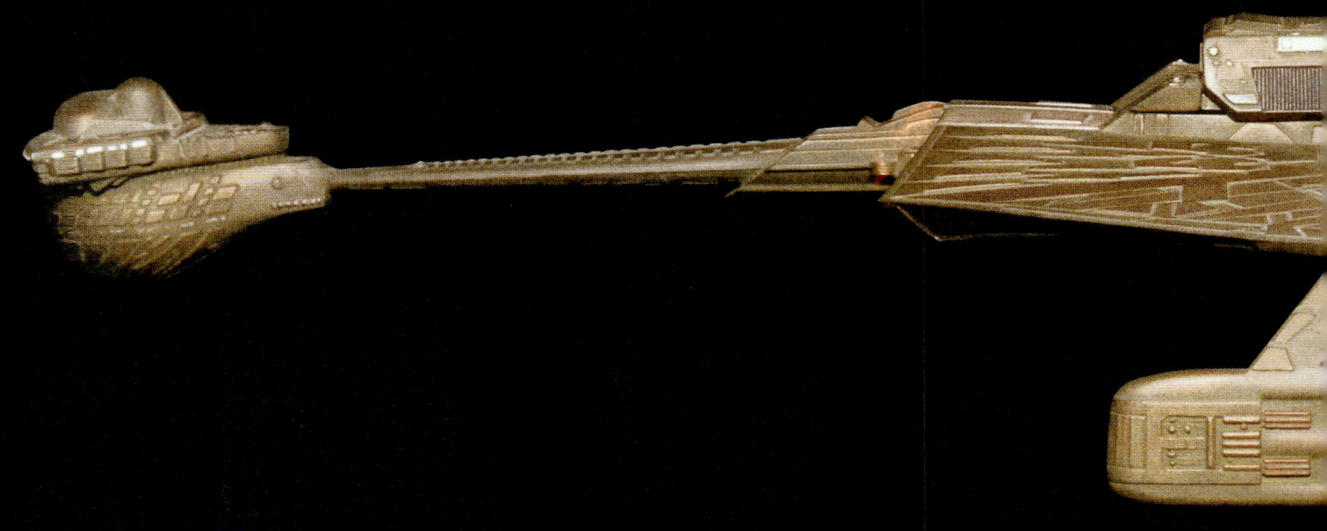

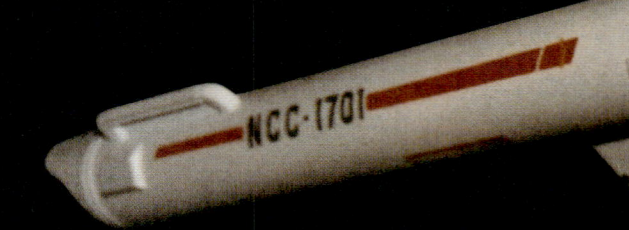

Top right: Klingon *K'vort*-class Bird of Prey from AMT/Ertl. Model by Rusty White

Above: Klingon D-7–class cruiser from AMT/Ertl. Model by Jerry Taylor

Bottom right: USS *Enterprise* NCC-1701 from AMT/Ertl. Model by Rick Jackson

Left: USS *Excelsior* from AMT/Ertl. Model by Rick Jackson

Bottom left: Kazon warship from Revell/Monogram. Model by James Staley

Bottom right: USS *Reliant* from AMT/Ertl. Model by Rusty White

Right: Klingon *Vor'cha*-class cruiser from AMT/Ertl. Model by Jerry Taylor

Below: Kazon Torpedo from Revell/Monogram. Model by Bob Jones

Bottom right: Cardassian *Galor*-class cruiser from AMT/Ertl. Model by Rick Jackson

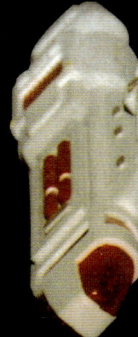

8
The Borg Cube

The Borg are a race of beings whose purpose is to improve themselves by "assimilating" other races of beings to benefit from their technology and culture. Introduced in the Next Gen episode "Q Who," they quickly became one of the greatest threats to the Federation because of both their intent and overwhelming power.

Creating a Borg ship is different from other projects. The simple appearance of a cube belies the texture and depth that must be added in order to gain the desired appearance. There isn't an "accurate" way to build a Borg ship—this is as much an artistic effort as anything.

Start with the basic cube. This can be constructed in any size and made out of a wide variety of materials. The factors to consider are size and weight of the model, inherent strength (i.e., does it need internal bracing?), and what glues you wish to use. A less significant factor is what, if any, texture you want on the outside of the base cube that will eventually become part of the texture and depth of the model.

Fig. 8-1. You can incorporate spare parts into the ship to add detail.

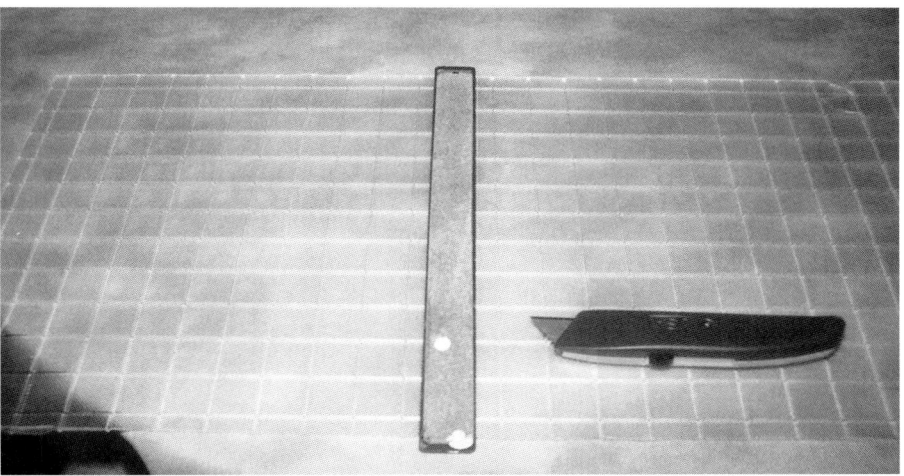

Fig. 8-2. Carefully score both sides of the plastic.

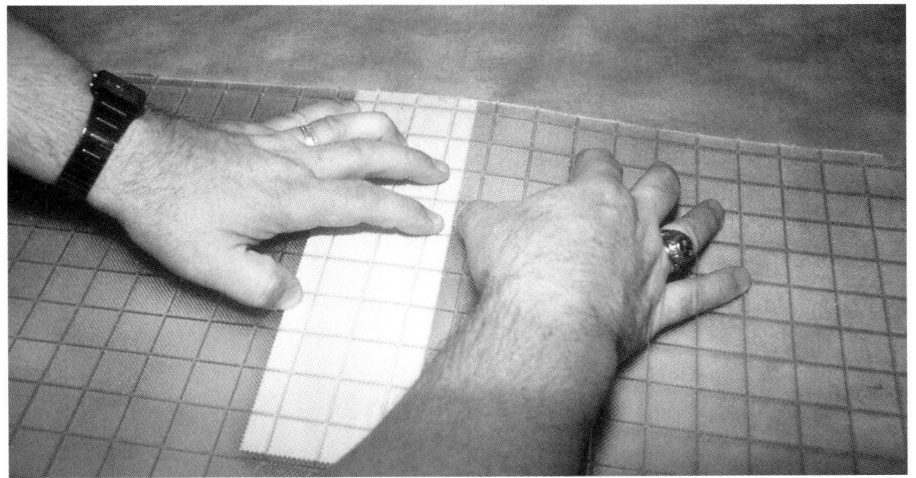

Fig. 8-3. Place it over a distinct edge such as a countertop or table to break the sheet at the scored line.

Paper-based products would be the lightest and least expensive. Choices such as corrugated cardboard, foamcore board, and Masonite would probably work best. Smaller cubes can be made from sheet styrene available at hobby shops, while larger cubes can be made using sheets purchased in bulk at plastic supply companies. Use at least .050 sheet for adequate stiffness.

Another alternative (and the one I chose) would be replacement lenses for 2 x 4-foot fluorescent light. I found one that had a grid pattern that I wanted to incorporate into the surface features.

One last thing to consider before buying. How do you want to attach the pieces? Dissimilar materials will limit the choices of glues when you're attaching the pieces directly to the cube.

Creating the surface detail

Most of the surface consists of pipes, girders, flat plates, and various odds and ends. Strip styrene, rods, and tubing of various sizes will provide most of the model. Floral wire in the 20 to 26 gauge range is a good source for the longer runs of piping. Throw in some other shapes such as I-beams, H-beams, C-sections, etc. Both Evergreen styrene products and Plastruct products have a wide range of shapes to choose from. Check with your local hobby or railroad shop. If they have a catalog, see what there is to order.

Look at railroad detailing parts for other possibilities, especially in the photoetched items like grilles. Photoetching can be expensive, so choose only what is essential.

Your parts box is also a source of inspiration. If you have been building for a while, you will have accumulated leftover parts from kits. You are especially looking for items that have additional detailing that would be difficult to scratchbuild (fig. 8-1). You'll want to attach some of these pieces directly to the cube, while others will look best mounted on one of the flat plates. You may also have leftover photo-etched parts that will look right. Don't ignore the possibility of using pieces from unbuilt kits. Taking a casting (see below) from a "virgin" part won't hurt it, and it may be just what you need to dress up the model.

Assembling the cube

Some of these materials are thin enough or soft enough to cut through cleanly. Others will have to be scored to start the process. The lens I chose was acrylic and brittle, so it falls into the latter category. Carefully score the plastic two to three times on the smooth side with a utility or plastic cutting knife. Turn it over and score on the other side one to two times. If it has ribs, be sure to score them as well (fig. 8-2).

Place the piece over an edge like a countertop or table, smooth side down. Align the scored line with the edge and gently press down on the overhanging part of the sheet (fig. 8-3). Don't try to go through completely, but listen for a crack and then move down an inch or two. After moving up and down the scored area in this way, you should get a clean break. Sand as necessary to get a flat edge.

Use a gluing trough or a picture frame holder set at 90° to hold the sheets while gluing. Reinforce the joint with some type of ribbing to eliminate flexing and to provide more surface area for the glue (fig. 8-4). Use Weld-On designed for acrylic plastic, since it will also work on styrene. It is available in types ranging from thin/quick-setting to gel/slow-setting. I chose no. 4 for the best combination of thinness and quick setup time.

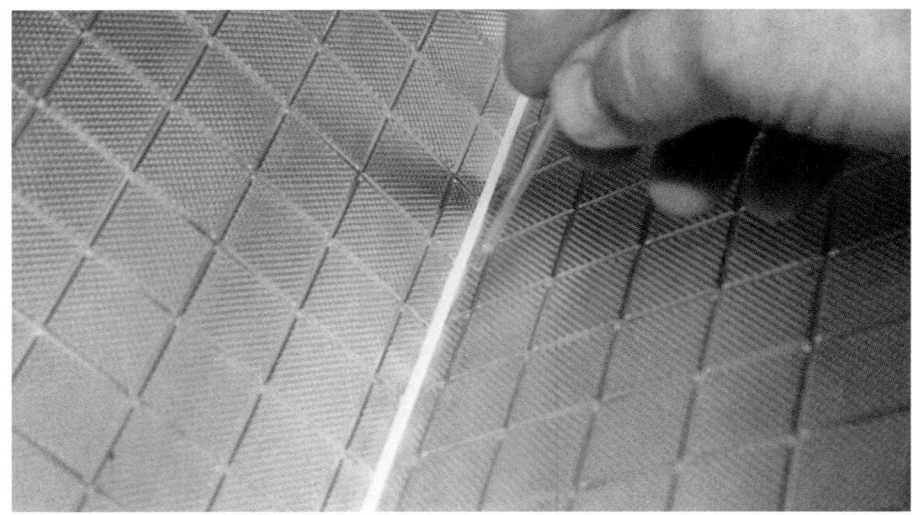

Fig. 8-4. Reinforce the joint with some type of ribbing to prevent the edge from flexing and to provide more surface area for the glue.

Fig. 8-5. I've epoxied 1/16" K&S brass strips to the inside of the sides. Again, note the styrene strips at the joints to provide more gluing surface.

Fig. 8-6. Also add ribs to the outside joints.

Fig. 8-7. Arrange the bracing for the top sheet in such a way as to distribute the stress evenly and prevent the mounting post from punching through.

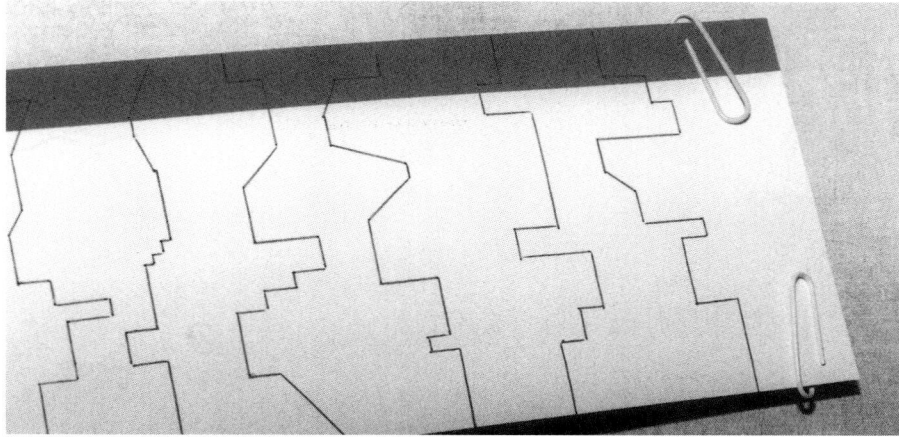

Fig. 8-8. Combining several sheets together allows you to produce the same pattern over and over.

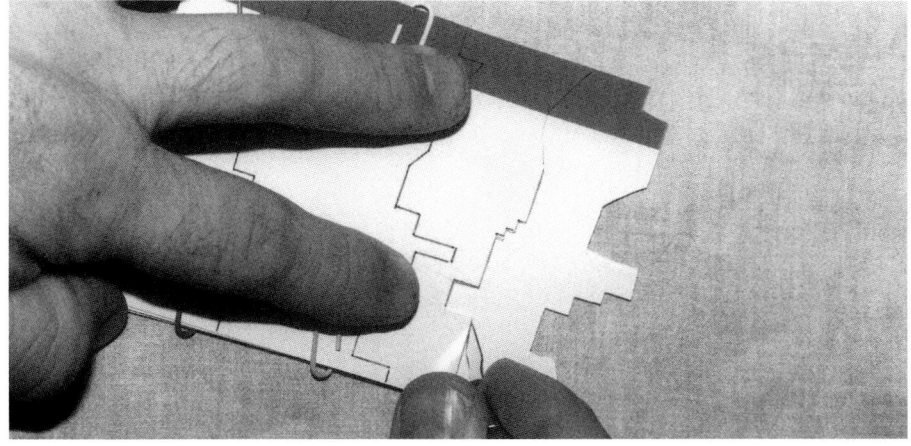

Fig. 8-9. Use the previous layers as a template for cutting out the lower layers.

A cube this size (12") needs internal cross-bracing to prevent the sides from bowing in the middle. One of the braces can double as the mounting post. I epoxied $1/16$" K&S brass strips to the inside of the sides (fig. 8-5).

You'll need to add ribs to the outside joints. Again, these add strength, but you are also providing attachment points for the surface detail (fig. 8-6).

The top is the last side of the cube to be attached. The center post to display the model is a K&S brass tube that will act as a sleeve. Drill a hole in the center of the bottom to slide the tube through. Test-fit the cube to make sure the tube is about 1 inch longer than the cube dimensions.

Attach the braces to the top sheet in a pattern that allows the post to hit them. Score the sheet around the contact point to improve the grip of the adhesive. Use two-part epoxy putty (found in the adhesive section of a hardware store) to attach the post to the braces in the top of the cube (fig. 8-7). This approach will distribute the stress of the model's weight evenly and prevent it from punching through.

Creating molds for casting masters

There are two ways to mass-produce the panels for the surface detail. The first is to clamp several sheets of stock styrene and trace patterns for the panels (fig. 8-8). Cut out each set of panels and add parts from the spares box to create the surface detail (fig. 8-9).

Rather than scratchbuilding everything, create a few simple masters and take molds from them using either latex or RTV (room temperature vulcanizing) silicon rubber molding material and casting resin. The latex should be available at any railroad hobby or craft store or it can be ordered.

RTV is more of a specialty item and might be available in your area only through a catalog. It also tends to be about four times more expensive than latex; however, production time for creating an RTV mold can be much faster (depending on the type you are using). So, if time is a factor, try to locate a source for RTV.

The masters should be simple. As you cast copies, you'll add detailing to dress them up and make them unique. By changing the position and orientation and adding details, you can use a small number of shapes over and over. Also, using molds, you can make copies of a small portion of a larger piece that, in total, won't work.

Be very careful when constructing the masters. The mold material is precise and will copy every aspect of the surface detail. Any imperfections not removed from the master will be duplicated in each casting.

If you're using latex to make the molds, attach the master to a piece of sheet styrene or other non-porous card. Use dissimilar glue, such as white glue, to attach plastic so the bond won't be permanent. Slowly paint on a thin layer of the latex, being sure to get the surface detail covered completely and removing any bubbles that might be in the latex (fig. 8-10). A thin layer dries in 3 to 8 hours. Allow 12 hours between layers if possible. As it dries, the latex will change from a milky white color to a translucent tan. Too thick a layer, and the milky appearance won't disappear. This means the side not exposed to the air (the side you want to cast) will still be soft when you separate the master from the mold, damaging the image.

Once the first layer has cured, add three to five additional layers until the mass of the mold is sufficient to retain its shape for casting. Large

Fig. 8-10. Apply the latex in thin layers. The latex turns from an off-white to a translucent tan as it dries. Once each layer has cured, you can add additional layers for strength.

Fig. 8-11. Pour the necessary amounts of the two RTV components into a mixing cup and stir until it's thoroughly mixed. Be careful not to create bubbles during mixing.

Fig. 8-12. Slowly pour the RTV into the former. Again, avoid creating bubbles.

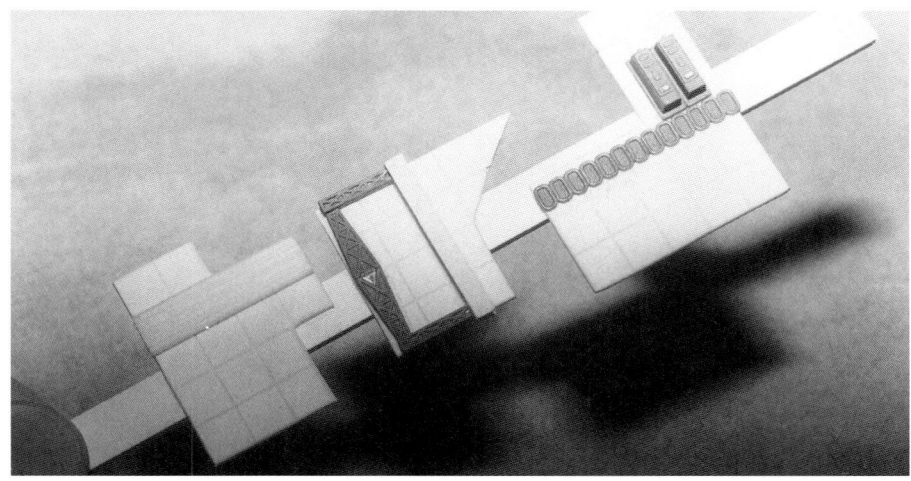

Fig. 8-13. Make masters of the plates by cutting various shapes from sheet styrene and attaching basic parts.

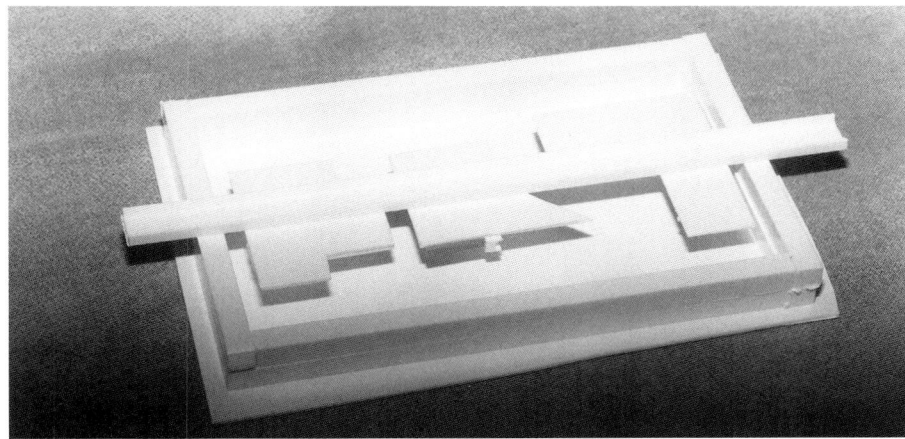

Fig. 8-14. Build a former to hold the RTV while it cures. Attach the masters to a bracket that rests on the former while the masters are suspended face down into the RTV.

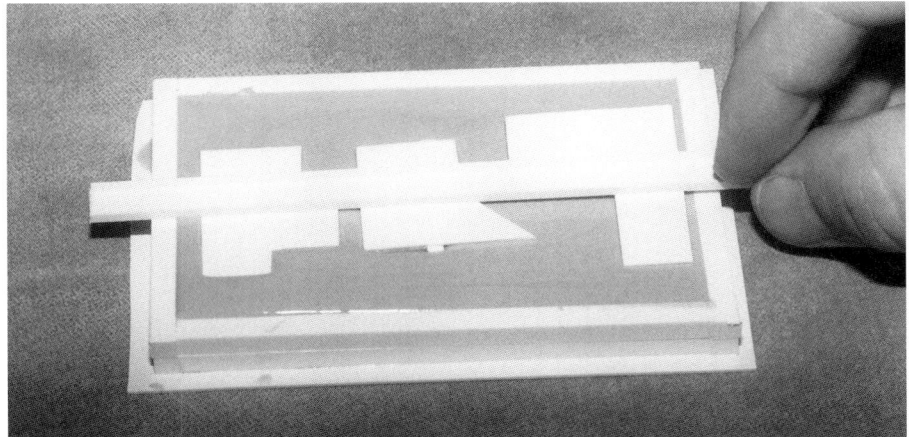

Fig. 8-15. Insert the master into the RTV. Let the RTV gradually flow around the parts so air isn't trapped in pockets on the underside.

molds may need some type of reinforcement, such as fibers or plastic strips imbedded in the later layers to prevent undesirable flexing.

RTV molds require making a "former" to contain the RTV while it cures. Be sure it is sufficiently deep and has sufficient border around the part. Plastic strips are usually the best materials to make the former, since the RTV won't stick to them.

Mix the RTV in the proportions recommended in the instructions (fig. 8-11). Mix only as much as you will need for the group of molds you're making. As you mix and pour the RTV into the former, be careful to avoid creating bubbles in the mold (fig. 8-12). Any trapped bubbles can settle against the surface of the master and eliminate the details. The best way to eliminate air bubbles is to use a vacuum pump to reduce the pressure around the parts and cause the bubbles to rise to the surface. Building a vacuum pump is outside the scope of this book.

The master will be suspended upside down in the surface of the RTV. To prevent the master from sinking into the RTV while it is curing, attach strips of styrene to the back of the master so they can rest on the edge of the former while the RTV cures (figs. 8-13, 8-14, and 8-15).

While both types of materials will work as described, you should try to use the RTV if at all possible. For this type of work, it is more durable than latex, so it will survive more castings. RTV has a natural slickness that performs the function of a mold-release agent. Despite its excellent reproductive characteristics, latex has a natural tack that increases the grip on the castings. The parts in latex molds are more difficult to remove without adding a release agent that might obliterate some of the surface detail. Not using a release agent runs the risk of tearing or distorting the latex mold while removing the part.

Casting the parts

Two-part casting resin can be found at hobby shops or ordered from catalogs. It ranges in viscosity from water-thin to syrupy. Thinner is generally better if you are using gravity to fill the mold. A thinner resin will also prevent bubbles from getting into the part.

Mix equal parts of the resin in a flexible, plastic working cup (fig. 8-16). Some resin is available as a kit with the cups included, but you can also use cups from nighttime cold medications. If at all possible, get a cup with a scale to measure by. As you stir, the two parts act like oil and water—there is a noticeable difference in appearance. Continue to stir until the difference disappears; otherwise, the mix will cure unevenly (fig. 8-17).

Another variable in the different brands of resin is the pot time. Some resins begin to catalyze in less than a minute while others may take three to four minutes. The more volume you have in a batch or part, the sooner the curing will start. When doing multiple pourings, slower and smaller batches are better. Regardless of the pot time, mix only what you can cast in one pass, because the excess will catalyze before you can make another run.

Pour the resin into the mold(s) (figs. 8-18 and 8-19). As it cures, it will change from a clear liquid to an opaque solid. If the proportions are right, the part should cure in less than 5 minutes, but the more massive parts will take up to 30 minutes. Let the part cure for at least double the time indicated, just in case the proportions were off or the mixing wasn't thorough. A part removed too soon will be

Fig. 8-16. Pour measured amounts of resin into two separate cups and combine as much as you need for a casting run into a third cup.

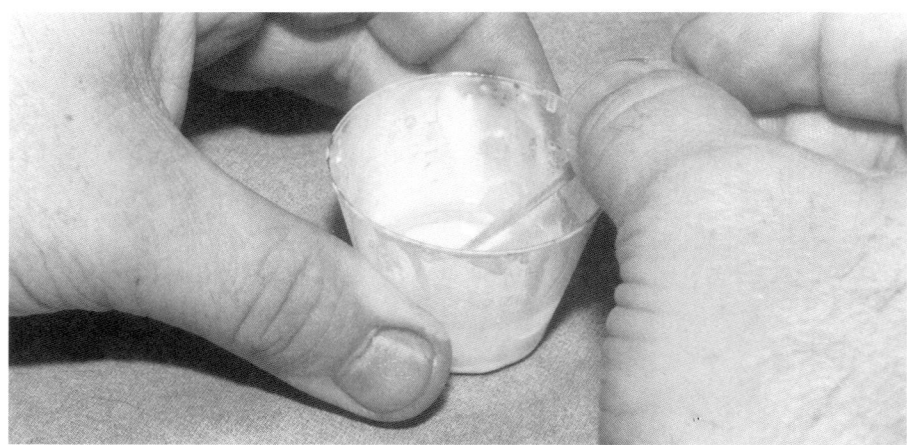

Fig. 8-17. Thoroughly mix the two parts of resin.

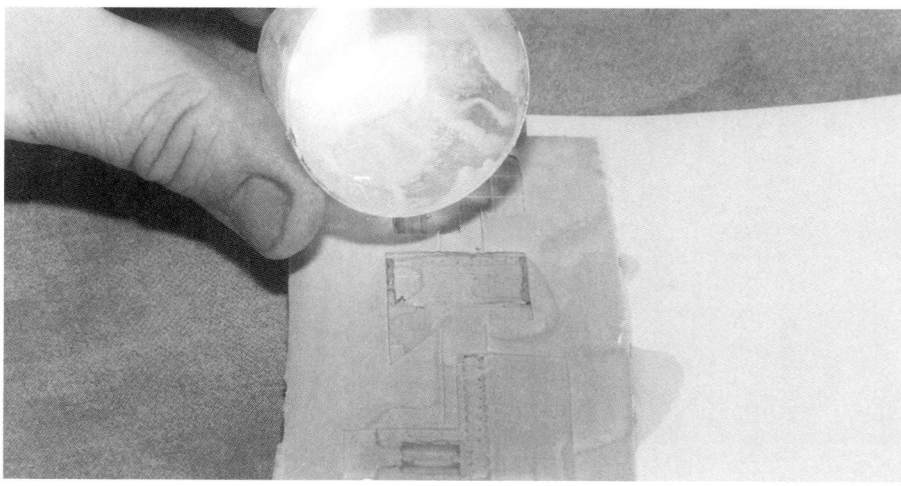

Fig. 8-18. Gently pour the resin into the mold. Work any bubbles out of the mix before it starts to set.

Fig. 8-19. As the resin begins to catalyze, it will turn opaque.

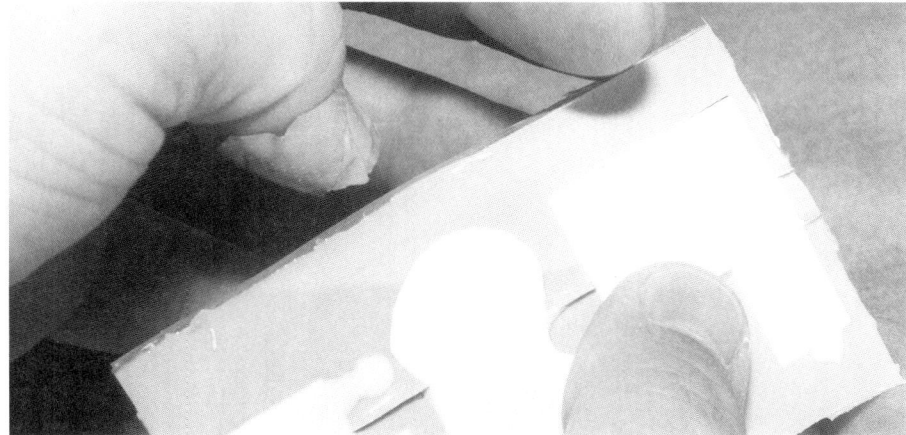

Fig. 8-20. Gently flex the mold to break the part loose.

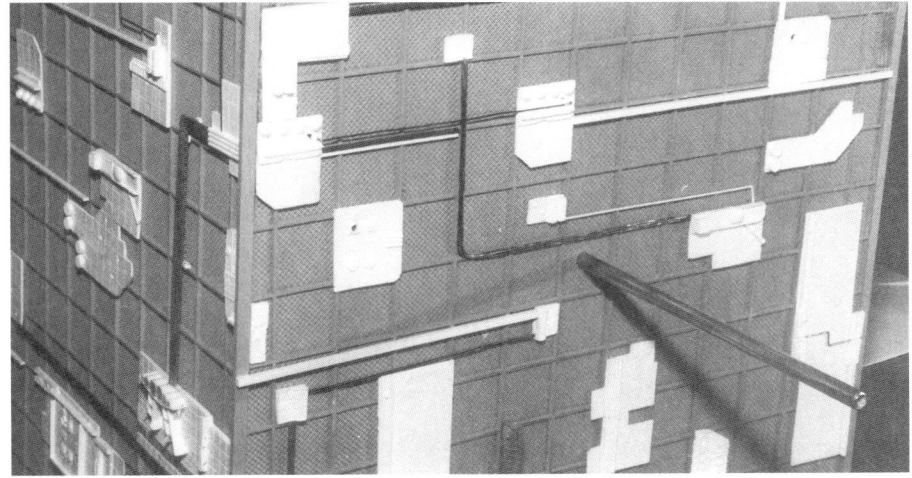

Fig. 8-21. Add the bottom layer of plates directly to the surface of the cube in whatever pattern seems appropriate.

soft and bend, making it useless. If the parts are no longer warm, it should be safe to remove them.

During the curing process, you will see the effect of any bubbles in the liquid. Curing creates heat and the heat causes the air to expand. Even the tiniest bubbles can become large enough to turn the piece into Swiss cheese. Again, the longer the pot time and the thinner the resin, the better the chance that the air will rise to near the surface of the mold, away from the casting detail. This is another instance where a vacuum pump can be used to eliminate bubbles.

Carefully flex the mold to remove the casting (fig. 8-20). Work the edges free first, then gradually go deeper and deeper into the mold. If the part is large and has a lot of surface detail that makes it harder to remove, sink a rod or piece of sprue into the back of the part to give yourself a handle to work the part free. It can be trimmed off later.

Attaching the surface detail

Inspect and clean up the castings before attaching them. Most will have some overrun edges and will be thicker than desired. Use a sanding block to clean up the edges and thin the part to a uniform thickness. Address any flaws in the casting at this time too. The lower layers can be a little rougher than the top layer, but be careful not to cut too many corners.

Build the surface detail in layers. Run one layer primarily in one direction around the cube. Then, build the next layer at 90° to the first. Alternate directions and use a minimum of three layers. Make sure that each layer is sufficiently "busy" but doesn't totally obscure the lower levels. Don't be afraid to cluster plates together to make larger shapes, or to trim away sections to make additional variations.

Add runs of piping and girderwork between the fixtures in each layer. Using floral wire in the 18 to 26 gauge range will provide a variety of looks. Create single and multiple runs as well as straight and dog-legs.

The basic plates you have cast will need "dressing up." This is where the artistic side is expressed in adding the extra strips, rods, and other pieces to break up the large, featureless areas.

Paint each layer before applying the next layer; otherwise, coverage will be difficult. Use a medium to dark gray (I chose Model Master Gunship Gray FS 36118) for the base coat and then highlight the layer with a lighter gray, such as Light Ghost Gray. Don't use too dark a color (especially on the bottom layers) or you will cause the texture to disappear. Finally, go in to add some details. A little bit of color and a variety of shades are appropriate. On the lower levels go a little overboard, since you will end up obscuring some of the color with subsequent layers.

You can attach the bottom layer to the cube and paint and weather everything at once. On subsequent layers, paint and weather the parts and subassemblies before attaching them, so you don't overpaint the lower layer.

Weather each layer by first applying a wash of India ink. Prepare a solution

Fig. 8-22. The finished Borg Cube. Perhaps it would be more appropriate to refer to it as *a* Borg Cube, since yours will undoubtedly look different.

of about 4 fluid ounces (about twice the size of a standard paint bottle) of denatured alcohol. Add 5 to 7 drops of ink to the alcohol and mix. Prepare a second container of straight alcohol.

You want to "overdo" the wash— apply it in much larger quantities than you would paint. Apply enough wash so it is able to run and puddle at the lowest level of the parts. As the alcohol carrier evaporates, the pigment will gather and bring out the shadow areas of the model. Because the ratio is very thin, you'll then find that the "over-do" results in the proper amount of shading.

Appendix 1: Paint

The appendices cover topics that are applicable to all of the chapters in this book. Rather than repeat a long explanation in each chapter, I am assuming (yes, I *know* the saying) that you will have read this material.

Federal Standard (FS) number

The Federal government uses a color identification system to classify paint by shade and intensity. The number identifies a paint by family (blues, greens, etc.), luster (flat, semi-gloss, gloss), and a numeric scale that proceeds from darker to lighter shades. For example, what is commonly known as Gunship Gray has an FS number of *36118*.
- *3* indicates a flat finish
- *6* is the gray color family
- *118* is a low number, indicating this is a relatively dark shade of gray.

Not every number has been used. As new colors are developed, they will be assigned a number between the colors they are similar to.

No matter what the manufacturer uses for a catalog number or what cute name he places on the bottle, all FS36118 paints should be the same color, shade, and finish. This does not mean that all gray paints are alike. As you go up the numeric scale from darker to lighter, tints will be seen. Some grays have a blue cast, while others (e.g. Gloss Gull Gray) have a yellow shading, even though they have nearly identical FS numbers.

The paint chips currently in use are available in publication FED-STD-595B (Volume 1) from the General Services Administration. The cost for the fan-fold version is $35; for the loose-leaf version, $40. To get one, send a check or money order payable to "GSA/Specifications Section" as well as a self-addressed label to the address listed in Sources. Some copies of FS 595A may still be available. The two systems are pretty consistent, so don't worry about not having both.

Aircraft modelers in particular use FS numbers to help select the paint for their models. Both paint and kit manufacturers have picked up on the idea by applying the FS number to many of their paints, making it easier for a modeler to determine which "dark gray" is appropriate.

The approach has been so successful that almost all newer kits, including the *Star Trek* kits, tend to use FS numbers to indicate the proper shade of paint.

Sometimes, however, you are asked to mix paints to achieve a color not provided by a paint manufacturer.

FS595A and FS595B provide accurate color chips used by the U.S. Government. Most model companies now use them as references for painting the models.

While mixing may achieve the most accurate shade, it can be difficult to predict how much paint you will need when making a batch. If you guess wrong while mixing a second batch, the color may be off a little because the proportions aren't exactly the same as in the first. If this is a concern but precision of shade isn't, select shades that are as close as you can get but can also be bought off the rack, thus relying on the manufacturer's quality control.

Scale effect of paint

Many modelers use a principle called "scale effect" when building their models. This states that paint intensity must be adjusted as you get smaller in scale. Any paint color on a 1:1 scale subject appears to lighten (fade) to a less intense color as you get farther away from the subject because of the intervening haze. Since this distance is simulated by the reduction in scale of the model, modelers add white or gray to each color of their paint. In this way, being three feet away from a model can simulate being 200 feet from the real thing.

Ignoring for a moment the lack of haze in outer space, you may want to adjust the starkness of the white portions of the models with some light gray. You may go so far as to substitute Camouflage Gray (FS 36622) for the white. To further complicate the issue, do you want to model the colors found on the studio model or the apparent colors found on your TV screen?

There isn't a right answer. It's a matter of taste. Look at what other modelers have done and experiment with the process yourself to decide which you prefer.

Appendix 2: Decals

Silvering

"Silvering" is the term used to describe the effect of light going through the clear portions of a decal when it does not conform to the surface of the model. When light enters at an angle on a properly applied decal, it goes through the decal and immediately strikes the paint on the model. It then bounces off at the same angle as it entered, resulting in a transparent appearance. Light enters a silvered decal the same way, but it strikes an uneven surface underneath. Instead of bouncing back at the proper angle, it begins to bounce around under the decal, against other portions of the uneven surface. This "pinball" action results in the shine that is the sign of a silvered decal.

The way to prevent silvered decals is to *always* have a glossy, smooth surface under the decals. It doesn't matter what the final finish is to be. If the decals don't have a smooth surface to lie on, you are asking for trouble!

One way to deal with silvering is to trim away all of the clear decal around the image, but this works only if the image is solid. Any clear areas inside (such as the interior of letters) still present a risk.

As you build the model, spend time trying to achieve as smooth a surface as possible. All decals need a glossy surface to avoid silvering. (It is not enough for a surface to be shiny, because even a rough surface can be shiny before it becomes glossy.) Use a clear lacquer such as Testors Glosscote to smooth out the surface. Depending on the surface, two to three light coats should achieve the necessary results. Another option is to use an acrylic overcoat. Many modelers like to use clear acrylic floor wax as the gloss coat, and they achieve great results.

Over the years, decal technology has improved significantly. Modern decals are much thinner and tougher. This makes them bend and stretch, so they can do a better job of following the surface shape. Decals in *Star Trek* kits, hard and thick in the early releases, have become extremely "modeler-friendly." In most cases, all you need to apply a perfect set of markings is water.

If your kit decals are the uncooperative kind, several manufacturers have developed liquids that will help. These products soften decals and allow them to conform snugly to the model's surface. Some are fairly mild, while others are extremely strong—so strong, in fact, that they can turn a decal into a puddle of goo.

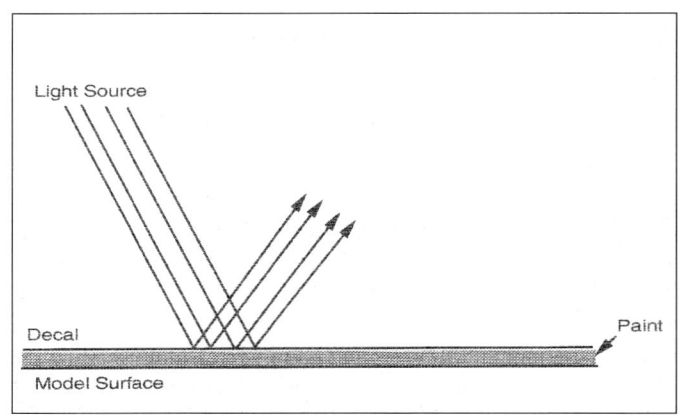

A properly prepared surface lets light pass through the decal without affecting the path.

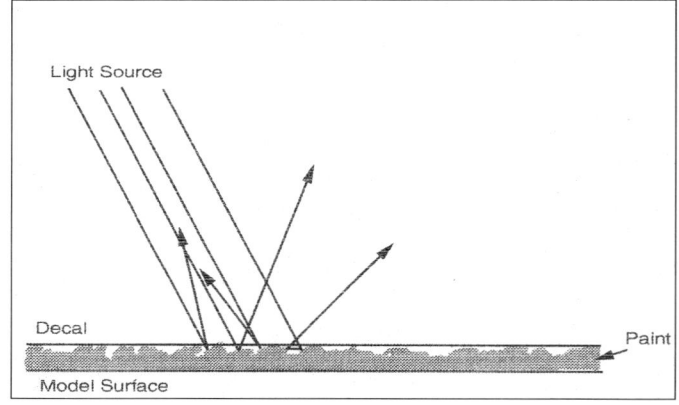

Flat paint has a rough surface that scatters the light that hits it.

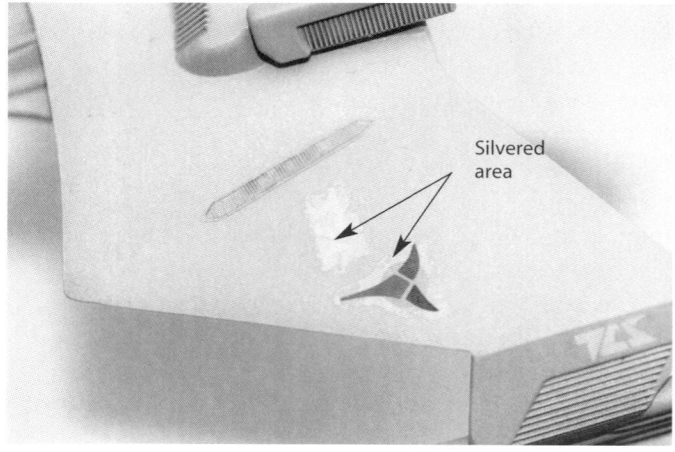

What happens when the surface isn't smooth and the decal isn't snug.

Appendix 3: Tools, Supplies, Sources

Building models requires practice; it also requires the right tools. Sometimes these are specifically designed for modeling. Most of the time, they are adaptations of everyday items. The photos below include items I find most useful.

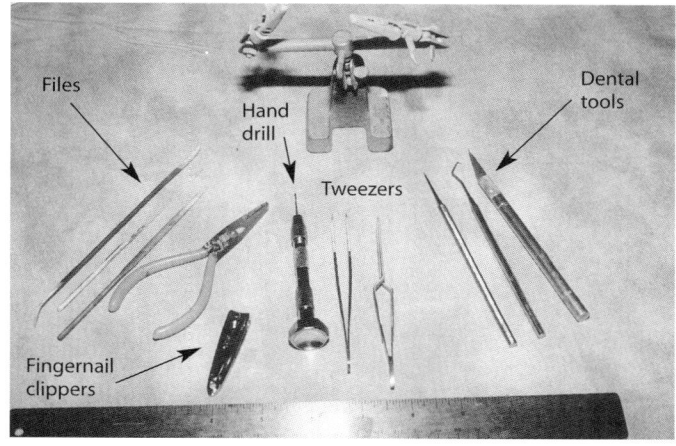

Most hand tools used for general modeling have the same application in the *Star Trek* universe.

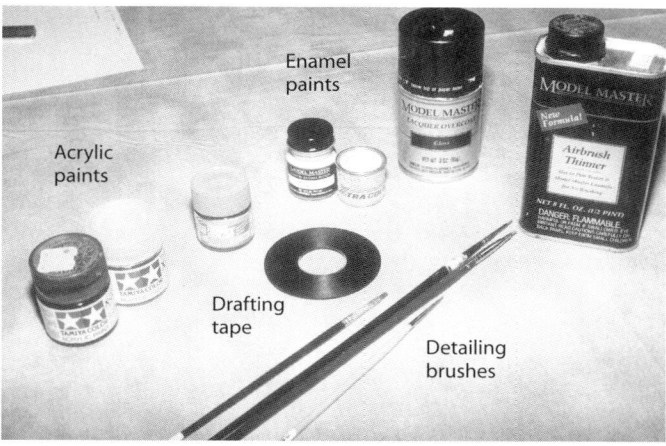

You'll need a wide range of paints. Detail brushes and masks will help with the complicated paint schemes.

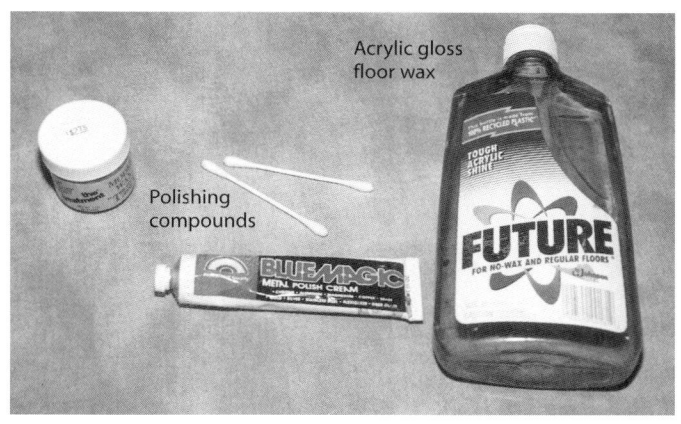

Various polishing compounds and overcoats can be used to achieve a smooth finish on the model.

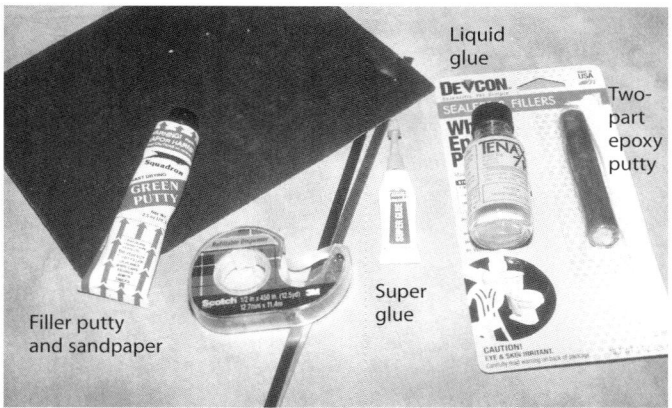

Fillers and adhesives are essential to eliminating the seams and producing an appealing model.

Kits

AMT and AMT/Ertl
The Ertl Company, Inc.
Highways 136 & 20
P.O. Box 500
Dyersville, IA 52040-0500

Revell/Monogram
8601 Waukegan Rd
Morton Grove, IL 60053

Sheet and strip styrene, tubing, and rods

Evergreen Scale Models
12808 NE 125th Way
Kirkland, WA 98034

Plastruct
1020 S. Wallace Place
City of Industry, CA 91748

Brass, steel, and copper products

K&S Engineering
6917 W. 59th St.
Chicago, IL 60638

Decals and decaling products

Web Games
240 Spicewood Lane
Hendersonville, NC 28791-1343

Super Scale International Inc.
Carson City, NV
(Setting solutions, Super Mask, Kristal Klear)

Moldmaking and casting materials, Kristal Klear

Micro-Mark
340 Snyder Ave.
Berkeley Heights, NJ 07922

ETI
300 South Bay Depot Rd.
Fields Landing, CA 95537

Paints

Humbrol
Marfleet Lane
Kingston-Upon-Hull
England HU9 5NE

Testors
620 Buckbee St.
Rockford, IL 61104

Gunze-Sangyo
Imported through
Marco Polo Import, Inc.
532 S Coralridge Pl.
City of Industry, CA 91746

Other building supplies

Blue Magic Polish Corp.
Center Street
Santa Ana, CA 92703
(also available at most automotive supply stores in the "Polishes" section)

IPS Corporation
455 West Victoria Street
Compton, CA 90220
(Weld-On)

Hebco
306 Briar Hollow Rd.
Hohenwald, TN 38462
(Tenax 7R)

Treatment Products Ltd.
4701 West Augusta Boulevard
Chicago, IL 60651
("The Treatment" polishing wax)

Creations Unlimited Hobby Products
4318 Plainfield Northeast
Comstock Park, MI 49321
(Touch-N-Flow applicator)

Squadron Products
1115 Crowley Dr.
Carrollton, TX 75011-5010
(Squadron Filler Putty)

References, information, and inspiration

The Collectors Value Guide for Scale Model Plastic Kits by John W. Burns

Star Trek: "Where No One Has Gone Before" by J. M. Dillard

Jackill's Star Fleet Reference Manual, Vol. 1–3 plus the Jackill's Technical Readout Data Sheets by Eric Kristiansen

*FED-STD-595B
Federal Supply Services
Specification Section
470 L'Enfant Plaza SW
Suite 8100
Washington DC, 20407
Web site:
http://pub.fss.gsa.gov/pub/fed-specs.html*

Numerous fan-based *Star Trek* web sites. Some sites are very stable while others appear and disappear overnight. Use a search engine to locate the current selections and URL addresses.